THREE STOPS
TO THE SUMMIT

Cover photograph:
Clogwyn Station
by Soichi Kaji

THREE STOPS TO THE SUMMIT

by

ROL WILLIAMS

Dedicated to the staff of the Snowdon Mountain Railway past and present.

"No one looking on Snowdon can deny that it is a mountain in its majesty, and that in form it is absolutely perfect."
S. Baring-Gould

First impression: May 1990

First published in Welsh as
Heibio Hebron
in November 1987.

ISBN 0 905775 84 8

Cover design and typesetting:
Stiwdio Mei

Printing:
Snowdonia Press (cover)
Argraffdy Arfon

Published by

CYHOEDDIADAU MEI
Pen-y-groes, Caernarfon,
Gwynedd.

CONTENTS

ACKNOWLEDGEMENTS

May I thank the following for the support and encouragement given in the preparation and publication of this book.

Derek Rogerson, Managing Director, and staff of the Snowdon Mountain Railway.

Staff of the Archives Department of the Gwynedd County Council.

John Roberts Williams for his advice and guidance.

The good people of Llanberis and the other villages for use of their photographs.

and Cyhoeddiadau Mei, Pen-y-groes, for once again 'getting the train on its tracks'.

Rol Williams,
April 1990

BIBLIOGRAPHY

BOYD, J.I.C. *Narrow Gauge Railways in North Caernarvonshire.*
JENKINS, D.E. *Folklore of Beddgelert.*
RANSOME-WALLIS, P. *Snowdon Mountain Railway.*
TURNER, Keith *Snowdon Mountain Railway.*

Caernarvon & Denbigh Herald
Gossiping Guide to Wales
Herald Cymraeg
Liverpool Daily Post
North Wales Chronicle
North Wales Observer & Express
Papur Pawb
Y Genedl

CHAPTER ONE

The Pioneers

Over the centuries the compelling summit of Snowdon has attracted innumerable travellers. Some have attempted the journey by a variety of unconventional and dangerous methods, riding motorbikes, bicycles, cars. One person even reached the peak on stilts. Most people, however, have chosen to follow the slow route on foot, pausing every now and then to rest and enjoy the spectacular views, whilst experienced climbers gain a sense of achievement by conquering the mountain by a far more dangerous route.

The first recorded climber to reach the summit was a botanist called Thomas Johnson, studying the various and sometimes rare flora and fauna to be found on the slopes of the mountain. The early Celtic people were the true conquerors of the mountains, and years before had stood valiantly on most of the Snowdon peaks.

Snowdonia became popular with tourists around the middle of the nineteenth century. This was presumably allied to the opening of the railway between Chester and North Wales and the new railway station at Bangor in 1848. This line was extended to Caernarfon in 1852 and a branch line was opened to the village of Llanberis in July 1869 by Sir Richard Moon, the then chairman of the London & North Western Railway. At one time there were plans to build a railway direct from Bangor to Llanberis, but though these plans were at an advanced stage, this idea did not materialize.

There is a reference to one Elias Roberts of Llanberis being carried on horseback to the summit in 1829. Forty years later, the same Elias Roberts went to the top with two horses pulling a cart full of pitch which was to be used to light a bonfire at the summit to commemorate the coming of age of the young heir to the Vaynol estate at Port Dinorwic (Y Felinheli), namely George William Duff Assheton Smith. Most of Snowdon and the land stretching from the summit to the shores of the Menai Strait belonged to the Vaynol estate and it could be said that George William Duff of all the squires of Vaynol had the greatest influence on the social, economic and cultural life of the small villages in and around Snowdonia.

A fairly recent innovation has been the annual race to the summit in July and the event has become so popular that various national and international runners attempt the gruelling five mile run to the top and back again to the Ddôl Fawr field at the lakeside. The record at present is one hour two minutes and twenty-two seconds recorded in 1985 and it is likely that some intrepid runner will eventually cover the distance in less than an hour.

To those who desired to be accompanied to the summit in the last century, the Snowdon guides would presumably be ideal partners. They were employed by the various hostelries in the area and some of them achieved great respect. In accompanying the various ladies and gentlemen, the guides would be prepared for any emergency that could arise *en route* and so they carried various alcoholic beverages to overcome any possibilities of their charges becoming weak, tired or disillusioned. The guides would accompany them from Llanberis, Penygwryd, Pen-y-pass, Beddgelert or Rhyd-ddu and well-known guides were synonymous with those centres — John Closs, Hugh Shone, William Wiliams at Llanberis, Robert Hughes and Henry Owen at Penygwryd, Thomas Jones at Beddgelert and Thomas Williams, or 'Caergyb' as he was universally known, at Rhyd-ddu.

Climbing Snowdon at any time is not only an arduous task but also a thirsty one. This was realised by some of the local entrepreneurs and around 1840 a stone hut was erected on the Nant Gwynant side of Snowdon by a gentleman called Morris Williams who originally came from the Amlwch district of Anglesey. He was employed in the copper mine at Clogwyn Coch and whilst following his employment on the mountain side, observed that a substantial number of people traversed the mountain footpaths during the summer months, people who could be looked upon as a potential buyers of various refreshments. Morris Williams was a monoglot Welshman and his inability to converse in English proved to be detrimental to any success in his proposed business venture. His only hope was to secure a suitable partner who could not only converse in English but was also familiar with local history. One such man was William Williams, who during his time had been a Snowdon guide.

In order to attract attention Williams was dressed in goat skin from head to foot and stood outside the hut, the subject of hilarity and admiration. Above the door of their little hut was a sign inviting all and sundry to 'Come and see the wild man of the mountain' and the naïve having been drawn to see the so-called 'wild man' invariably bought teas and sandwiches! Business boomed and in time Morris Williams sold his share in the venture to his brother Phillip, who later secured as his partner a certain John Roberts of Llanberis. They erected another hut near the summit, which at that time was owned by three different landlords, namely the Hafod y Llan estate (Sir Edward Watkin) of Hafod y Llan, Nant Gwynant, the Baron Hill estate, Beaumaris, Anglesey (Sir Richard Williams-Bulkely), and of course the Vaynol estate (George William Duff Assheton Smith).

Phillip Williams and John Roberts felt that it would be beneficial to extend their refreshment venture by obtaining a licence to sell alcoholic drinks in their hut and for some unknown reason the application was submitted in the name of John Roberts only. The magistrates at Caernarfon agreed to the application, but having obtained the licence, John Roberts refused to allow his partner to benefit from such a concession. Naturally the partnership between Williams and Roberts was dissolved. Phillip realised that the summit of Snowdon was actually in two parishes and so, since part of the land was in the parish of Beddgelert, he applied to the magistrate at Porthmadog for a licence. After being unsuccessful on more than one occasion, he eventually succeeded and it is reported that for some years two licences were granted to sell alcohol at the summit — one licence granted by the Porthmadog bench and the other by the Caernarfon justices.

Wil Williams had returned to his duties as a Snowdon guide and was known locally as 'Wil Boots' as he was employed at the Dolbadarn and Victoria hotels, where one of his duties was cleaning visitors' boots. Wil was renowned as an expert on Snowdonian flora and frequently requested to accompany the gentry on biological surveys in Snowdonia. It was on one such occasion that Wil was requested to obtain a sample of a rare fern that grew on a ledge above Llyn Du'r Arddu. He had tied a rope to his body and was lowered on to the ledge. The rope snapped and poor Wil fell to his death. On his tombstone in the churchyard at Nant Peris are carved the following words:

UNDERNEATH LIE THE REMAINS
OF WILLIAM WILLIAMS
UPWARD OF TWENTY FIVE
YEARS BOTANICAL GUIDE
AT THE ROYAL VICTORIA
HOTEL WHO WAS KILLED
BY A FALL
JUNE 13 1861, WHILST
PURSUING HIS FAVOURITE
VOCATION

This tombstone was erected to his memory by a few friends.

The part of Snowdon where he fell to his death is known in the mountaineering fraternity as 'Wil Boots gully' and at one time a piece of land by the lakeside at Llanberis was referred to as 'Dôl Wil Boots'.

CHAPTER TWO

Initial Endeavours

For many years various influential persons had been discussing the possibility of laying a railway track to the summit of Snowdon. In November 1871 a Bill was presented in Parliament to incorporate a company that intended to build such a railway, in the first place by purchasing the land and then building stations, bridges and other buildings. The land belonged to George William Duff, the squire of Vaynol, who was the elder son of Robert George Duff and his wife Mary, the niece of Thomas Assheton Smith of Vaynol who had been responsible for building the imposing Victoria Hotel at Llanberis. George William Duff had succeeded to the estate in 1859 when only eleven years of age, as a result of his aunt's will. His property extended into twenty-six parishes and he could, if desired, walk from Vaynol to the top of Snowdon without once stepping off his own property! As a token of esteem for his great aunt's generosity, George added her surname, Assheton Smith, to his own and so the squires of Vaynol for years were known as Duff Assheton Smith. In 1888, George married Laura Alice Stanhope Jones of Cheshire (and on his marriage remitted half their rent to his 1,500 tenants). The last squire of the Vaynol estate, Sir Michael Duff, relinquished the 'Assheton Smith' from his name and informed all and sundry that he desired to be known as Sir Michael Duff.

Without doubt the influence of the Vaynol family on the lives of their tenants, who toiled on their small-holdings on the slopes of the Snowdonian hills or worked in the huge Dinorwig Quarry, was significant and far-reaching. Thus, any company thinking of erecting a railway to the top of Snowdon had to obtain the blessing and agreement of the squire of the land, and this was not forthcoming.

During the final thirty years of the nineteenth century, railways had been built in various parts of Britain and this was reflected in the numerous measures presented to Parliament. The Snowdon Railway Bill was discussed in the House of Lords on Tuesday, 13 February 1872. Hence the description of that measure:

> . . . an Act for making a railway from Llanberis in the county of Carnarvon to the summit of Snowdon, in the same county and for other purposes . . .

A select Committee of five Lords discussed the Bill in detail.

This venture was estimated to cost £20,000 – £15,000 of which was to be spent on the railway and the remaining £5,000 on bridges, roads, walls, fencing, detonating rocks and field drainage.

G.W. Duff Assheton Smith

The preparation proved fruitless, however, and the Bill was withdrawn. Mr Assheton Smith would not give it his support and he was indeed against any railways on Snowdon, since he felt that it would not only scar the landscape, but it would not, in his opinion, bring any tangible benefit to the village of Llanberis. In passing, it should be borne in mind that he agreed in 1895 to build a railway line to transport slates from the Dinorwig Quarry to Penscoins at Port Dinorwic. (Part of this line is today used as a tourist attraction and known as the Lake Railway.) Assheton Smith felt that such an asset was enough of a contribution to the well-being of the area.

Following the opening of the LNWR branch line from Caernarfon to Llanberis in 1869 (later the LMS line), local poets mused on the possibility of Llanberis becoming a tourist area. Edward Foulkes of Erw Fair composed two stanzas in strict metre to glorify the occasion and indeed Sir Richard Moon, chairman of the London and North Western Railway, when opening the Llanberis branch line stated that it was the desire of his company to build a railway line all the way from London to the very summit of Snowdon itself! Bearing in mind such intent one is bound to ask why the offices and station of the mountain railway were not built adjacent to the mainline station as had been the case with the North Wales Narrow Gauge Railway (later the Welsh Highland Railway) at nearby Dinas, Llanwnda, which was enjoying much success? The intention was to run a train from Dinas station all the way to Porthmadog, some twenty-three miles through unequalled scenery along the rivers Gwyrfai and Glaslyn and passing the villages of Rhostryfan, Waunfawr, Betws Garmon, Rhyd-ddu and Beddgelert. Indeed the line had been laid as far as Llyn Cwellyn by mid 1878 and by the spring of 1881 had reached the village of Rhyd-ddu — four years after the first rail had been laid at Dinas, Llanwnda.

Naturally, the inhabitants of Beddgelert were determined that their village should become the main centre of tourism and mountaineering in Snowdonia, with ample and commendable hostelries to cater for and accommodate the visitors. Indeed there were rumours that plans were afoot to extend the railway from Dolwyddelan by means of various tunnels through the mountains to Nant Gwynant or else to have the Betws-y-coed branch extended to Beddgelert via the Gwynant valley. One should not ignore the fact that Sir Edward Watkin was the owner of the Hafod y Llan estate in Nant Gwynant at that time and his considerable expertise and technical know-how with railways would certainly have been beneficial to any local company hoping to pursue the intention of glorifying Beddgelert as the true and only centre of Snowdonian tourism. He had been a former chairman of the Manchester, Lincolnshire and Sheffield railway and had also been one of the pioneers of a cross-Channel railway tunnel to the continent.

The prime minister of the day, William Ewart Gladstone, had spent a week-end in Nant Gwynant in September 1892 as a guest of Sir Edward and Lady Watkin. The purpose of the visit was to open officially to the public the path known as the 'Watkin Path' to the top of Snowdon.

During the same period leading figures in the public life of Llanberis had been constantly visiting Vaynol to try and convince Mr Assheton Smith that allowing a railway line to be built on the slopes of Snowdon would be of immeasurable benefit to the whole area. On each and every visit the landowner refused to accept their prophecies. On June 1894 there appeared an editorial article in the Welsh language weekly newspaper *Yr Herald Cymraeg* stating that there was local discontent that the area was devoid of local improvements and that some concern had been expressed by local businessmen at the lack of any positive endeavour to attract visitors to the area. A local committee was set up with W.H. Jones, Liverpool House, as secretary and Richard Owen, a local printer, as chairman with the sole purpose of improving local amenities. (Captain Stewart, agent to the Vaynol estate, in company with the local rent collector, had visited all the smallholders who occupied the multitude of smallholdings on the slopes of Snowdon, every one stating that they had no oppostion to the building of a railway across their land.)

The North Wales Narrow Gauge Railway changed the name of the Rhyd-ddu station to Snowdon Station (South Snowdon at one time) — certainly a misnomer since the station was a few miles from Snowdon, but if the action was to attract people on to the railway, it served its purpose as the visitors poured into the area by means of what the company's booklet called 'the shortest and more picturesque route to Snowdon. Visitors should bear in mind that in climbing Snowdon by this line (ours) they are conveyed by the "toy" railway to a height of nearly 900 feet above sea-level so that they have quite two miles to traverse from Snowdon station, compared to the Llanberis distance and about 800 feet less to reach the summit'.

CHAPTER THREE

The Company is Formed

In 1893 another deputation from Llanberis went to see Assheton Smith at Vaynol and impressed upon the landowner and squire the fact that the village of Beddgelert was surely well on the way to being accepted as the focal point of tourism in Snowdonia, and warned him that unless some positive and remedial action were taken immediately Llanberis would lose out. Having implored Assheton Smith to allow the building of a railway they left to return home. Within a few days Assheton Smith gave his decision. Yes, he would allow a company to be set up to build a railway from Llanberis to the summit of Snowdon. In November 1894 the new company, The Snowdon Mountain Tramroad and Hotels Co Ltd, was formed and the first meeting of the directors was held at the Queens Hotel, Chester. On behalf of the company, Colonel Cragg from Kent agreed to the purchase of the requisite land from Assheton Smith who also agreed to lease the Royal Victoria Hotel for a period of fifty years from 12 November 1894, the annual rent to be £220 with a right of the new company to terminate the lease after five years if they so desired. Thirty acres of land in and around the hotel, including land on which Dolbadarn castle stood, were also leased to the company — the hotel had received royal patronage after the visit of Princess Victoria to Llanberis in 1832. The railway company paid £1,500 for the land together with the transfer of 6% of the shares which were sold on the market at £10 each.

The *Caernarvon & Denbigh Herald* of 24 May 1895 stated

that the new company would have a capital of £70,000 with 6,343 shares on the market, in addition to debentures of £100 each. The seven directors consisted of members of aristocratic families in Anglesey and Caernarvonshire, many of whom were retired Army officers with a sprinkling of business people. The seven were:

John Sutherland Harmood Banner (Chairman), (Accountant) Liverpool.

Harry Clegg, Plas Llanfair, Anglesey.

Colonel W.W. Cragg, Shortlands, Kent.

R.E.L. Naylor, Liverpool (Banker).

Lord Alexander Paget, Plas Newydd, Anglesey.

Major F.W. Turner, Plas Brereton, Caernarfon.

Captain N.P. Stewart, Bryn Tirion, Port Dinorwic, Vaynol Estate Agent

Mr Assheton Smith did not wish to be a director himself, but made a specific request that Captain Stewart should be a convenor of the company and a director. Sir Douglas Fox, a distinguished engineer from London, was appointed as Advisory Engineer to the project with his brother Francis Fox as a consultant. Messrs A.H. Holme and C.W. King of Liverpool were to be the main contractors, and the tender for the trackwork was allocated to Messrs Cammell of Sheffield and the local engineer appointed was Frank Oswell. Since the original idea was to electrify the line at some future date, Assheton Smith agreed that the company could utilize the waters of the Ceunant Mawr waterfall nearby should 'it be determined to supplement steam by electric traction'.

The general opinion was that the problems facing the new company were quite daunting. The land was extremely rocky and steep, and there existed no natural shelter should the weather be inclement, as it quite often was, on the bleak mountain terrain. Furthermore, the necessary machinery would have to be carried by horse-drawn sledges and this would be extremely difficult. In addition, it was necessary to build five stations, a number of bridges, and more of a

THE SUBSCRIPTION LIST WILL OPEN THURSDAY, MAY 23rd, AND CLOSE ON OR BEFORE THE FOLLOWING WEDNESDAY, MAY 29th.

THE

Snowdon Mountain Tramroad and Hotels Company

LIMITED.

(Incorporated under the Companies Acts, 1862 to 1890)

CAPITAL - - - - £70,000,

Divided into 7,000 Shares of £10 each.

Issue of 6,343 Shares of £10 each—and £20,000, part of an issue of £35,000 Four and a Half per Cent. First Mortgage Debentures of £100 each.

The Debentures will constitute a first charge upon the entire undertaking and property of the Company. They will be for 20 years, but redeemable at the option of the Company after the 1st July, 1905, upon 6 months' notice.

The Interest on the Debentures will be payable Half-Yearly, on 1st January and 1st July in each year.

FIRST MORTGAGE DEBENTURES

Payable £10 per £100 on Application.
£40 „ „ Allotment.
£50 „ „ 1st Jan., 1896.

SHARES

Payable 10/- per Share on Application.
£1 10/- „ „ Allotment.
£3 „ „ 1st Aug., 1895.
Balance in calls not exceeding £2 per Share and at intervals of not less than two months.

Allottees may, however, pay up in full on Allotment, Dividend and Interest being payable on the Shares and Debentures from dates of payment.

DIRECTORS.

J. S. HARMOOD BANNER, Esq., 24, North John Street, Liverpool, Chartered Accountant.
HARRY CLEGG, Esq., Plas Llanfair, Anglesea.
Lt-Col. WALLACE W. CRAGG, Southolms, Shortlands, Kent.
R. E. L. NAYLOR, Esq., 17, King Street, Liverpool, Banker.
Lord ALEXANDER PAGET, Plas Newydd, Anglesea.
Capt. N. P. STEWART, Bryntirion, Port Dinorwic.
F. W. TURNER, Esq., Plas Brereton, Carnarvon.

BANKERS.

LEYLAND & BULLINS, 11, Dale Street, Liverpool.
WILLIAMS & CO., Old Bank, Chester, and their Branches, and as their Agents—
ROBARTS, LUBBOCK & CO., 15, Lombard Street, London, E.C.

Solicitors—WALTER WEBB & CO., 23, Queen Victoria Street, London, E.C.

Engineers—Sir DOUGLAS FOX & FRANCIS FOX, 28, Victoria Street, Westminster, S.W.

Architects—F. & G. HOLME, Crosshall Street, Liverpool.

Secretary and Manager—G. C. AITCHISON.

Registered Office—VICTORIA HOTEL, LLANBERIS.

Temporary London Office—28, VICTORIA STREET, WESTMINSTER, S.W.

PROSPECTUS.

This Company was formed in November, 1894, for the purpose of:—

The prospectus of the new company.

challenge than any of the other tasks was the need to erect two viaducts over the river Hwch only a few yards from the base station at Llanberis.

Sir Douglas Fox and Gowrie Colquhoun Aitchinson, the newly appointed Secretary/Manager to the S.M.T. & H. Co Ltd visited Winterthur, near Zurich in Switzerland, to inspect the unique locomotives built by the Swiss Locomotive and Machine Company for the Swiss mountain railways. On their return they presented their report to the directors and unanimously suggested that one system was more acceptable than any of the others — this was a system devised by a Swiss engineer called Dr Roman S. Abt. The system, referred to as 'the Abt system' meant using a single steel bar with the teeth mechanically cut to the upper part of the bar. By having two bars side by side and having the teeth alternate to each other on the bars, it was possible to have a constant connection between the pinions on the train and the teeth on the railrack below, giving a smooth unbroken drive.

The company stated its intentions in the Prospectus:

(a) To build and work a tramroad its length approximately 4¾ miles from Llanberis to the summit of Snowdon.
(b) Erect, furnish and maintain a hotel at the summit of Snowdon at the terminus of the tramroad 3500 feet above sea level.
(c) Obtain lease, and maintain the renowned Victoria Hotel at Llanberis — situated near the starting point — and carry out a number of improvements to the hotel.
(d) Complete the laying of the tramroad so that the public may use it during the present season.
(N.B. It did not stipulate what was meant by 'the present season'!)
(e) Commence the tramroad at a point opposite the Victoria Hotel and near the terminus of the L.N.W.R. railway in Llanberis.
It is hoped that there will be two stations in the middle and a further station at the conclusion of the journey to the summit.

The last half mile to the summit of Snowdon.

It was further agreed that the railway company would have all the fishing rights of Llyn Peris, the upper lake.

Some consolation may have been derived from statistics given for a similar venture in Switzerland. The Rigi company had a capital of £50,000 and, during 1893, 45,000 passengers had been carried on their mountain line. The Snowdon line directors naturally hoped for a similar success.

In the contract made with Holme & King it was expected that the contractors would 'on or before the 1st day of July 1895 deliver over the Tramroad and works to the Company, fit to be opened for public traffic for passengers and goods' — a period of seven months from 15 November, the date the agreement was signed! Without doubt this was an impossible request bearing in mind the geological and geographical difficulties that existed. It was estimated that the total cost would be in the region of £64,000, consisting of £26,000 for clearing and preparing the land, £17,973 for the rail-track, £1,500 for legal costs, £12,000 for machinery, including locomotives, £1,575 for a signalling system, £1,152 allocated for advertisements and the paltry sum of £100 'as compensation to the farmers involved'. The remainder, £3,500, was allocated for any losses that might be incurred. The rail-track allocation included £300 for an engine shed 'to accommodate three engines and carriages with an ashpit costing £20 at the entrance, plus space to carry out repairs'. It was also intended to spend £1,200 on refurbishing the Victoria Hotel and buying furniture for the proposed hotel to be built on the summit.

The record books belonging to Gowrie Aitchison revealed that the proposed train fare in the first year of operation was to be three shillings — with a recommendation by Mr Aitchison that the wage of the guards should be six shillings a day, with the maintenance men and porters to receive five shillings. In the same book there was reference to a ton of sleepers costing £7 10s 0d, and a ton of fishplates costing £10. The gauge of the track was to be the same as most European lines of the same type, two feet, seven and a half inches (800mm).

The Abt System. Note the alternating teeth.

An agreement with the contractors, Holme & King, stipulated that they would receive 6,400 shares for completion of the work — these payments to be made periodically as part of the work was completed, but no payment was to be made without the approval of Sir Douglas Fox himself. There was a further stipulation that Holme & King were, at their own expense, to take every precaution whilst the work was in progress, to prevent any trouble or unruly behaviour by the workers and that the contractors should be responsible for employing special constables or some other form of policing, if so desired by the local magistrates.

The men lived and slept in wooden huts specifically erected as accommodation, premises which the contractors were expected 'to arrange for each workman in the huts provided by the contractors, to be provided with a separate bed which shall not be occupied during his absence'. It was also expected that no work was to be carried out on Sundays and no machinery to be moved on the Sabbath unless it was absolutely necessary.

Following the visit of Sir Douglas Fox and Gowrie Aitchison to Winterthur, when it was agreed to adopt the Abt system at Llanberis, an order was placed with the Swiss company for two locomotives, estimated to cost £1,400 each. An order for four carriages was made at the same time with the Lancaster Carriage and Wagon Company in Lancashire — they had been designed by Sir Douglas himself after he had seen those in use in Winterthur.

The work of erecting the two viaducts over the river Hwch was allocated to Chambers of Manchester and it was agreed that this work was to take precedence over all other aspects of the project. If possible the work was to be carried out during the winter months of 1894-95 so that the track itself could be laid during the spring of 1895. The intention was to erect the two viaducts of stone, with each arch — eighteen in all — to be of bricks. Without doubt, building the two viaducts was the major engineering work involved in the whole project — the main and lower viaduct was to be 166 yards long and the upper one of 63 yards, with each arch being four feet higher than the next. The only local company involved was one from Caernarfon, Owen Morris, who secured the contract for building the proposed stone walls on each side of the railway. (The same Owen Morris and his workmen had also been responsible for building the imposing extension to one of the local chapels, Capel Coch, in 1893.)

Travellers on the summit of Snowdon in times gone by with the guides and horses.

The Viaduct

CHAPTER FOUR

Letters to the Press

It is only fair to record that there was unstinted and general support to the venture from the inhabitants of Llanberis and other neighbouring villages. On the other hand there were some, more often than not from outside the area, who condemned the idea in no uncertain manner, one more so than the rest namely Canon H.O. Rawnsley, who in addition to being an Anglican priest was also the honorary secretary to the National Trust for the Preservation of Historic and Natural Beauty.

On 19 October 1894 a letter from the reverend gentleman, the first of many, appeared in the *North Wales Chronicle* (his address at the time being 'Tyn Ffynnon, Barmouth').

Dear Sir,

As Hon. Sec of the National Trust for the Preservation of Historic Interest and Natural Beauty you will, I trust, pardon me in writing to ask if there is any truth in the report that Mr Assheton Smith, the owner of part of Snowdon, is about to sanction the making of a railway to the top of the mountain and the building of a hotel there. It is quite possible that the rumour is without foundation. Meanwhile a large number of people in all parts of the United Kingdom are alarmed by the paragraphs that have appeared in the press releating to it. The deplorable example set on Snowdon will touch every mountain height.

Yours,
H.O. Rawnsley

Bearing in mind the extensive publicity which the scheme was attracting in both local and national press it was indeed surprising that the Canon was not better informed of what was contemplated in Llanberis.

At that time Captain Stewart, the estate agent, and one of the company directors replied to the letter.

23 October, 1894

Dear Sir,

The rumour you refer to about a railway to the summit of Snowdon is not without some foundation, for the line has been surveyed. I am not, however, in a position to say to a certainty that it will be carried out, at least, at present.

Two different Bills have been presented in Parliament at different times for this project, which Mr Assheton Smith on both occasions successfully opposed at very considerable expense. Still he feels that it is difficult to stay the inevitable and that sooner or later he must gracefully give way. He has now withdrawn opposition. He is not promoting the scheme himself but is granting reasonable facilities to others. Don't you think on reflection, however, that the view you take on the matter is rather selfish? Why should Mr Assheton Smith be debarred from promoting the interests and prosperity of the people amongst whom he dwells?

Why should Snowdon be reserved exclusively for the enjoyment of mountain climbers? Why should they have the entire monopoly of the mountain? Are there not thousands and tens of thousands of people, some too young and some too old and others, who from various causes find themselves unable to make the ascent, who would like to inhale the exhilarating air of the mountain and from the highest summit in England or Wales look down on the glorious panorama that spreads beneath?

Yours,

N.P Stewart.

To: Canon Rawnsley

23 October, 1894

To Captain Stewart

Thank you for your courteous reply. You lead me to believe that Mr Assheton Smith, to promote the best

FORM OF APPLICATION

FOR

FOUR AND A HALF PER CENT. FIRST MORTGAGE DEBENTURES OF £100 EACH.

THE

Snowdon Mountain Tramroad & Hotels Company

LIMITED.

To the Directors of

THE SNOWDON MOUNTAIN TRAMROAD AND HOTELS COMPANY LIMITED.

GENTLEMEN,

Having paid to your Bankers the sum of .... Pounds, being a deposit of £10 per cent. on my application for .... Pounds Four and a Half per Cent. First Mortgage Debentures of £100 each of the above Company, I hereby request that you will allot to me that amount of Debentures upon the terms of the Company's Prospectus, dated 21st May, 1895, and I agree to accept such Debentures, or any less amount that may be allotted to me, and to make the remaining payments in respect thereof at the dates specified in the said Prospectus.

Name (in full) ....

Address ....

Description ....

Date .... *1895.*

Signature ....

I desire to pay in full on allotment ....

THE SNOWDON MOUNTAIN TRAMROAD AND HOTELS COMPANY LIMITED.

BANKERS' RECEIPT.

Received, *this* .... *day of* *1895,*

of .... *the sum of*

.... *Pounds, being a Deposit of £10 per cent. on*

application for £ .... *Four and a Half per Cent. First Mortgage Debentures in the above Company.*

This Form to be sent entire, with a Deposit of £10 per cent. on the amount of Debentures applied for, to Messrs. Leyland & Bullins, 11, Dale Street, Liverpool; Messrs. Williams & Co., Old Bank, Chester, and Branches; or to Messrs. Robarts, Lubbock & Co., 15, Lombard Street, London, E.C.

An application form for debentures.

interests and prosperity of those among whom dwells, has at last been persuaded gracefully to give way, and from the tourist point of view, and for other reasons, to withdraw opposition to this objectionable scheme of vulgarising one of our grandest national possessions. Future England will deplore the day that allowed man, who are custodians of it to look upon mountain scenery chiefly or only as grist to the mill. Snowdon can now be mounted easily by all but the most infirm and Snowdon in a sense belongs not to Llanberis only, but to the world.

Yours truly,
H.O. Rawnsley.

At the end of October, Captain Stewart replied to the Canon's letter hoping no doubt that the brevity of his reply would put an end to any further discussion!

Dear Sir,

I have been away from home, or would have written to you sooner. I am glad you acknowledge my letter to have been a courteous one and I regret to discover so little of that quality in yours. When you find yourself defeated in argument you indulge in sentiment and I can only regard your epistle as the sticky sentimental dribble (canonised) of a dreamer and a faddist. I must ask you not to trouble me with any further communications.

Yours truly,
N.P. Stewart

But the redoubtable Canon was not to be silenced, for on 23 October 1894 he sent a letter to Assheton Smith himself.

Dear Sir,

I write now — in the name of the National Trust for the Preservation of Sites of Historic Interest and National Beauty to ask if, at this eleventh hour, you would, in the best interests of the nation, reconsider the matter; and if it could be shown that the prepondent sense of the nation were against such innovation, you would then take firm steps to prevent so sacred an inheritance as Snowdon being thus robbed of its chief charm for future generations, and being vulgarised for ever. It is in very few places in our crowded country that men can be alone with nature and with their

God — and Snowdon is one of them. To rob Snowdon, so easily accesible as it is both by night and day, of its grand national solitude and super-eminent charm, will be to inflict a loss upon the whole world.

Yours obediently,
H.O. Rawnsley

Mr Assheton Smith replied on the day of receipt.

Dear Sir,

In reply to your letter of this morning I regret that I cannot take the same view of the matter as your association appears to do.

You are right in saying that I was in former years opposed to the scheme, but times have changed, and if in many ways one does not advance with them, one is left alone in trying to direct the tourists to Llanberis, and making things easy for them. I am consulting the interest of this estate and the neighbourhood in which I live and I cannot recognise any outside interference in the matter.

Faithfully yours,
G.W. Duff Assheton Smith.

Though Captain Stewart had emphatically stated at the end of October that he was not going to proceed with any further correspondence it was rather surprising to see a further letter from him in *The North Wales Chronicle* on 8 November 1894.

Sir.

The Canon's poetic rhapsody about the 'future education of the hearts and eyes of England and Wales' may be very fine, and his flights of fancy most vivid, but his facts are bad and so is his judgement and his arguments are unsound. For instance, he contends that as Llanberis advances in prosperity, Beddgelert and Capel Curig must decline and become impoverished. There never was a greater mistake. Assuming that the railway will bring 1000 to the summit for every 100 persons who make the ascent now I venture to predict that out of every 1000, 500 at least will not return to Llanberis, but descend to Beddgelert and Capel Curig, thus bringing a steam of prosperity such as they never before enjoyed. Before the end of three years the good people of

Beddgelert and Capel Curig, to say nothing of Llanberis, will be showering blessings on the heads of the Snowdon railway promoters.

Then Canon Rawnsley goes on to say "It is in very few places in our crowded country that man can be alone with nature and with God and Snowdon is one of them". This observation shows how utterly ignorant he must be of the mountain. He cannot be aware that 500 a day is not a very unusual number to make the ascent, and I am informed that upwards of 1000 people climbed the mountain one day last summer. To talk of "the solitude" of Snowdon is bosh.

Yours,
N.P Stewart

The editor of the *North Wales Chronicle* presented his own comments as a postscript to the letter.

Canon Rawnsley, in protesting against the converting of the mountain into 'a mixture of a tea garden and a switch-back', urges that it is now easily accesible from Capel Curig, Beddgelert and Llanberis, and that ample pony tracks go within a few yards of the summit, and points out that although Llanberis may benefit from the railway, it will be at the expense of other villages from which the ascent is now made. His most serious objection, however, lies in the assertion that in order to be a commercial success, the mineral resources of the mountain will have to be exploited to an extent that will destroy the natural beauty of the spot.

The London Star in November 1894, under the heading 'Smith-ing Snowdon' published a very critical article in which it referred, so it seemed, to the manner in which the Vaynol estate had become owners of Snowdon and the surrounding lands.

This article infuriated Assheton Smith and on 7 February 1895 the following apology appeared in the edition of the *London Star*:

We desire to express our regret having published on 6th November 1894 this article and to unreservedely withdraw the charges made against you.

We are satisfied that the withdrawal of your opposition to

the railway was only made in deference to the wishes of your neighbours and tenants, and that you were not actuated by motives of pecuniary gain to yourself.

We are equally satisfied that the charges made that you and your fore-fathers had appropriated thousands of acres from the common were also without foundation and we admit that you have done nothing in the management of your estates or your dealings with your tenants to justify the assertion that you are what the Irish call a 'land grabber'. We do not admit that you rightly interpreted our article, or that there were any intentions on our part to make these charges against you, but we recognise that you have a right to complain, and beg to tender you our sincere apologies for having published the article.

It seemed that the Canon was in the minority and support for the proposed railway came from within and outside Wales. A certain R.J. Corbet from Shrewsbury wrote to *The North Wales Chronicle* at the end of November 1894.

Having known the Snowdon district for 30 years, I am well aware how much the great mountain could be increased by giving 'the old man in a hurry', the middle aged one and the young facilities for ascending to heights, sublime without the laborious walk of 5 miles from base to summit. No pedestrian would be robbed of his walk by the making of a funicular or other railway, and only one side of the giant hill would be disfigured . . . I feel confident that neither a cultured man like Canon Rawnsley nor more homely lovers of grand scenery need fear anything from the multiplication of excursionists, male and female, learned and ignorant, serious and frivolous.

On November 1894 Dr W.O. Lloyd Williams, the local doctor at Llanberis and a very revered gentleman in the locality, wrote to *The Manchester Guardian* and his letter showed the intense support in Llanberis for the proposed railway.

Sir,

Poor Llanberis! for my friends your genial and cultured writer of the "Welsh Notes" is tempted to leave us to fate.

And why? Because we lack the patriotism to resist the making of a railway up Snowdon. It is true that we favour the railway. Some time ago Mr Assheton Smith, through his agent, communicated to our parish authorities the intention of the promoters of the railway and requested them to ascertain the views of the inhabitant in the matter. At a well attended public meeting, called by the improvement committee for another purpose, the above communication was announced and received with much enthusiasm, and a resolution was passed unanimously promising every facility the inhabitant could afford to expedite the project.

It is certainly our opinion that the railway will make Llanberis more attractive to visitors and will give them an easy access to the summit, from which they will enjoy the grandest and most varied scenery in our country. It is absurd to say, as Canon Rawnsley does, that none but the absolutely infirm need fear the ascent of Snowdon; the journey up and down is very exhausting except to the strong and healthy and most of those who cannot walk or are either unfit, or too timid to ride. It would appear that your correspondent does not relish the idea of mixing on the sacred mountain with what he calls "the bun and whisky tourists" by which uncomplimentary name, I presume, he means the working classes, who are sometimes tempted by cheap fares to enjoy a day in the fresh and exhilarating air of our mountain. Although my experience — limited, I admit — of the fares charged on mountain railways does not lead me to anticipate much of that class of traffic, I cannot agree with your correspondent that it would be in the least degree deplorable to attract to the summit of Snowdon as many as can by persuaded to avail themselves of so healthy a pastime. If your correspondent and Canon Rawnsley — for we hope to provide for them both — require more reserve than they could secure under such circumstances, we have plenty of accommodation in the surrounding great mountains and even on Snowdon itself, where they will not be disturbed by so much as the echo of the 'latest comic song'.

I am
yours
W.O. Lloyd Williams.

The L.A.D.A.S. *loco'. The driver's name was William Pickles.*
(Photograph by courtesy of Ken Jones.)

Some of the local authorities also supported the idea, one of those being the Tourist Committee at Llandudno.

> . . . We have not the least hesitation in saying that myriads of visitors to Llandudno, who are physically incapable of ascending Snowdon will hail the proposed railway as a boon and blessing. Those who wish to toil up can still do so, and the young lady who last summer surprised a whole party by descending in about an hour can continue her muscular Christianity. But for goodness sake, let mortals with less muscular limbs have a chance of seeing nature from such an altitude.

The newspapers in general were most vociferous in their support. The editorial comment of *The North Wales Chronicle* in November 1894 was a shot in the arm for the new company and acclaimed Assheton Smith as a philanthropist worthy of all praise heaped upon him.

> The action of Mr Assheton Smith in granting facilities to the promoters of the proposed Snowdon Railway has called forth the indignant protest of the secretary of the Society for the Preservation of National Monuments, and the usual aesthetic persons always ready to join in a newspaper correspondence, which may serve to exhibit their 'superiority'. The position of Mr Assheton Smith in the matter was plainly indicated on the series of letters which were published in our last issue, and Canon Rawnsley, the Secretary of the Society above-named, had all the facts before him when he first lifted up his voice against the so-called desecration of Snowdon, but these facts seemed to have no effect whatever upon the rev. gentleman, whose sense of beauty is apparently out of all proportion to what ordinary folks would call his common-sense. It is always an easy matter to pose as a champion of the preservation of natural beauty, but the very fact that the sympathy of unthinking people is enlisted with ease should be a caution to plain men to be quite sure that there is really a good case before any question of desecration or vandalism is publicly raised. We venture to think that if Canon Rawnsley had tempered his platitudes with a little respect for common-sense and a regard for facts, he would not have failed to appreciate the kindness which prompted Mr Assheton Smith and his advisers to

grant facilities for the construction of the proposed Snowdon Railway. The high reputation which Mr Assheton Smith enjoys as a landlord in North Wales is well-known to our readers, so that it is necessary for us to say a word with reference to his gentle attitude towards his tenantry. It is sufficient to say that his action in the matter of the Snowdon Railway was taken in persuance of his general line of conduct, with sole regard for interests of the thousands of people who have to find their livelihood upon his estate. Mr Assheton Smith, personally, will not benefit one farthing by the construction of the Snowdon Railway, but it is hoped that the inhabitants of the neighbourhood around Llanberis will find that the facilities afforded by him will be the means of inducing larger number of tourists to attempt the ascent of Snowdon from their district. We have said that Mr Assheton Smith had regard only for the people dependent upon him when he agreed to withdraw his opposition to the Snowdon scheme. It is to be added that the generous and high-minded Squire of the Vaynol really sacrificed his prejudices in order to meet the pressing and fervent prayers of the people of Llanberis. Had Mr Assheton Smith turned a deaf ear to the petitions of his people the Radical and Socialist Press, which is now filled with shrieking denounications of the desecration of Snowdon, would not have ceased from holding up to obloquy the landlord who preferred his own sentimental fancy to the prosperity of the thousands who inhabit Llanberis. The people who live in the Snowdon district have no need to be told that it is no unusual sight in summer to see long trains crowded with tourists going in the direction of Snowdon, but it is equally well known that, by a clever device of the promoters of the Narrow Gauge Railway, the stream of visitors has been almost entirely diverted from Llanberis. The re-christening of the Rhyd-ddu terminus and calling it 'Snowdon' attracts thousands of tourists annually, with loss to lodging-house keepers, hotel-keepers and the general population of Llanberis..As Captain Stewart pointed out in one of his letters, it is absurd folly to speak of the solitudes of Snowdon during the summer months and it is equally silly to describe the people who make the acsent ac 'bun and whisky' tourists. As a matter of fact the crowds that visit Snowdon are orderly and decent people and if the new line brings such people to the neighbourhood in still larger numbers no one will have to complain, whilst the inhabitants of Llanberis will have the satisfaction of having a share of the money which is now spent in other parts of the Snowdon district, which are by no means as favourable for making ascent of the Welsh Monarch, but which, by clever advertising have been made popular at the expense of the quarry town.

Canon Rawnsley's efforts seemed to be in vain. The directors carried on with their plans to hold the ceremony to cut the first sod at the riverside at Tŷ Clwb not far from the grounds of the Victoria Hotel. It had been originally suggested that the ceremony would be on 5 December 1894 but on second thoughts it was realised that much more publicity and a greater crowd would be present if the ceremony was held on a Saturday. This seems to be confirmed in a telegram sent on 20 November:

From Capt Stewart to Holme & King
Assheton Smith fixed fifteenth December for sod cutting.
Stewart

From Holme & King to Capt Stewart
December 15 will suit me for sod cutting.
Holme

Llanberis and its inhabitants were ready for the historical day.

CHAPTER FIVE

Cutting the First Sod

Though an important election was being contested in Llanberis on Saturday, 15 December 1894, this did not deter from the arrangements made for the sod-cutting ceremony. There were flags and bunting everywhere and though the weather was rather cold with snow on most of the mountain tops, a large crowd had gathered by the riverside and on the main road to witness the ceremony at one o'clock and to welcome the Assheton Smiths and the directors of the new company.

The Llanberis Subscription Band, conducted by T.H. Williams, led the procession through the village, rendering Welsh airs to the delight of the crowd. Behind the band were several local dignitaries and representaives of various organisations, and at the back of the procession was the car containing G.W. Duff Assheton Smith and his six year old daughter, Enid. The ceremony was to be performed by Mrs Assheton Smith but as she had been indisposed for some weeks and unable to attend, the actual cutting was to be carried out by the young Enid.

The arrival of the procession at Tŷ Clwb was a signal for blasting to commence at the nearby Dinorwig Quarry (also owned by Assheton Smith). A selected few were invited to sit at the side of the small stage on which the squire and his daughter sat. Sir Douglas Fox invited young Enid to come forward to cut the sod and according to newspaper reports 'with deft hands and sparkling eyes, the young lady plunged the silver bladed spade into the grass'. The barrow into which the turf was put was of mahogany with a silver plate on its side commemorating the event.

> First sod of the Snowdon tramroad cut *December 15th 1894* by Miss Assheton Smith.

Unfortunately, the present whereabouts of the spade and mahogany barrow are not known. Neither is it possible to locate the model of the wheelbarrow and the spade which were presented to young Enid.

Whilst the official ceremony was going on, there was a moment of comical distraction for the crowd. The cape of a certain Mr Kingsley, a Caernarfon photographer officially appointed to record the historical event, fell into the river and he ordered his young assistant to wade into the cold waters of Afon Hwch to retrieve it — according to some bystanders the young lad was far from pleased but received the adulation of the crowd.

Local opinion was expressed by W.H. Jones, Liverpool House and Thomas Hughes of Hafod Lydan whose address complimented Assheton Smith on his support.

> On behalf of the people of Llanberis we extend a very warm welcome to you on this very important day. We hoped to welcome Mrs Assheton Smith to the ceremony. With deep regret we are given to understand that her health prevented her from coming here today. We wish to thank you, Sir, for the interest shown by you in the success of the locality by allowing the venture to proceed. (Signed by W. Lloyd Williams, W.H. Jones, Thomas Hughes, G.H. Peale, Edward Foulkes, Richard Hughes. Dec. 15th 1894.)

Mr Assheton Smith in his reply thanked the gentlemen for their loyal addresses. He sincerely hoped that the railway would bring success to Llanberis by re-kindling tourism in the area but he had to give a word of warning: 'The hotels should not under any circumstance charge exhorbitant prices on any tourist who visited Llanberis or they would not visit the area again.'

Further remarks thanking the directors for their efforts to bring prosperity to Llanberis were made by Edward Foulkes

The hostelry on the summit. The notice above the door proclaims 'Bazaar and Refreshments. Well-aired beds. Ham and eggs. Choice beverages. Roberts & Owen, proprietors'.
(Photograph by courtesy of the Gwynedd Archives Service.)

of Erw Fair. Lord Alexander Paget replied on behalf of the directors and confirmed, to appreciative applause, that impending tourists would not be overcharged. 'The accommodation proposed and the charge to convey people to the top of Snowdon would be such that all tourists would recommend their friends to visit the area.'

The directors had given *carte blanche* to a well known Liverpool caterer, Phillip Eberle, to prepare a luncheon at the nearby Victoria Hotel, which incidentally had been newly-painted on the outside for the occasion for the princely sum of £56 5s 0d! Two hundred guests sat down to a meal which according to reports was 'as near an approach to perfection as could be desired' as will be noted from the following menu.

POTAGE
Tortue Claire

POISSON
Saumon du Rhin en Aspic a la Prince de Galles

ENTREES
Vol au Vent à la Financière
Cotelettes de Mouton puree de Pois
Salmi de Becassine à la Lucullus

ROTIS etc etc
Faisans Bécasse
Pâté de foie gras en croûte de Strasburg

ENTREMETS
Pudding Snowdon Gelée de Champagne
Crème à l'Italienne Dames d'honneurs

DESSERT
Bombe glacée Ananas Raisins
Poires Bananas etc etc

Should you visit the Royal Victoria Hotel today, you can if you so desire still taste the celebrated 'Pudding Snowdon'!

Amongst the guests were the Bishop of Bangor, Mr Grant Duff and Mr Harry Duff of Vaynol, Alderman and Mrs D.P. Williams, Llanberis, Mr and Mrs Brinkman, manager of the Dinorwig Quarry, Dr R.H. Mills Roberts and Dr W. Lloyd Williams, the local practitioners, Reverend J.O. Jones representing the local clergy, all directors and officials of the company.

The Lord Paget requested all present to join him in expressing ' "our deep regret and sorrow at Mrs Assheton Smith's enforced absence, alas, laid low on a bed of sickness. I ask you to drink the health, the restoration to health of Mrs Assheton Smith and that she may be speedily restored to that state of condolence which was the earnest wish of all." (Hear, hear.)'

A shoe from the hoof of one of the horses used to pull sledges up the mountain slopes. A horseshoe of average size is shown beneath for comparison. It weighed three and a half pounds and measured seven inches across.

(Photograph by courtesy of Mrs M. Bryn Hughes.)

Responding, Mr Assheton Smith thanked all present for the kind manner in which he had been received. ' "When I return to the mansion at Vaynol I will inform my wife of the admirable manner in which young Enid has carried out her duties in the absence of her mother." (Loud cheers.)' He stated that though he had been against the railway at the beginning he firmly believed that there would be a railway up Snowdon sooner or later, if not on this side, then it would be on the other side, for apparently there was a determined effort to divert the tourist traffic to the other side of the mountain. That, said Mr Assheton Smith, was to be prevented if possible. (A direct reference to the project of the Narrow Gauge Railway from Dinas to Beddgelert which by then had reached the village of Rhyd-ddu.)

He then referred to a letter he had received from Canon Rawnsley whom he believed to be a member of the Society called 'The Mountain Protection Society'. ' "This was a private letter," he said, "and I was not aware that it was going further than that, but somehow it got into the press. Our correspondence has ended but I have heard that the reverend gentlemen is still firing away from a distance. I have heard from a gentlemen who knows Canon Rawnsley and that he is a gentleman who changes his views. He lives in the neighbourhood of Keswick, and when the Manchester Corporation broached its scheme for appropriating Thirlmere, Canon Rawnsley was one of its most determined opponents. Now, strange to say, when the scheme has matured he was one of the first to congratulate the Manchester Corporation on the success of the undertaking. The directors of the Snowdon railway might therefore, hope that they would yet have the distinguished pleasure of carrying Canon Rawnsley to the top of Snowdon." (Applause.)'

Sir Douglas Fox in referring to the 'great advantage of having the owner of the land with us', also complimented Captain Stewart, 'who had watched the matter from the begining to the end and care has been taken to deal in the most tender way with all the beautiful scenery, especially the waterfall and glen which leads to it.'

The local bards again rushed into song, and many of the local Welsh language newspapers of the period published their poetical endeavours to the delight of the inhabitants of Llanberis and other villages.

At the end of 1894 the slate quarrying industry was at a rather low ebb, but it received a local boost with a proposal not only to build a mountain railway, but also by an announcement by a construction company that it intended to build 150 houses at Llanberis and had already purchased land for this purpose.

Llanberis, indeed, was waking up!

CHAPTER SIX

The Work Progresses

A progress report in February 1895 showed that the inclement weather of the winter months had delayed much of the work and concern was expressed that it would not be completed by the first day of July 1895 as agreed. Nevertheless, excavating work had been carried out near the Ceunant Mawr waterfall where traces of copper had been found. The contractors, Holme & King, were still optimistic that they would complete the work as contracted and that trains would be running in June. The extent of the wintry conditions can be gauged by a newspaper reference at the time to the fact that the whole of Llyn Padarn had frozen over in February and that Mr Rigby, one of the railway engineers, had been skiing on the lake on the 22nd of the month 'with 20 fathoms of water under him'!

Whilst shop assistants in Llanberis itself were voicing their opinions on the streets and later in the court for a weekly half-day holiday, by mid March 1895 one hundred and fifty men were employed on the railway scheme and, in spite of the weather, work on the two viaducts had progressed favourably. Sir Douglas Fox was hopeful that as soon as the weather improved, the company would be employing a further two hundred and fifty men. Two and a half miles of track had been cleared by the end of March in spite of the elements and the extremely bleak terrain.

It was hoped that the first locomotive would arrive from Switzerland by the summer months and its arrival would be a great boost. The relationship between the contractors and the farming community, whose stock grazed the land on both sides of the proposed track, was cordial and Sir Douglas had insisted at the beginning that the smallholders should not be impeded in any way from carrying out their duties and special places would be erected so that they could take their animals across the line. It was intended that the men would live and sleep in wooden huts erected at the side of the line, and would be allowed to go wherever they pleased at week-ends — more often than not to the local hostelries who certainly enjoyed their patronage. Naturally the construction of the hotel on the summit had to be delayed until the track had been laid to the top when the materials could be conveyed by train. One reporter of the local *The Caernarvon & Denbigh Herald* was certainly impressed by what he saw when visiting Llanberis in early 1895.

> Our correspondent then took leave convinced that when Mr Oswell (one of the engineers) and his merry men shall have conquered Snowdon they will be competent to proceed at once to subdue Popactapetl or even the mountains of the moon!

By April 1895 two hundred men were employed and again there were constant reports of the delay caused by the harsh winter months. But even then the contractors were still hopeful of completing the work by the contract date — the first of July! Some three miles of track-bed had been prepared 'ready for the rail' and the foundations of four bridges were ready for concreting at the end of May.

The appointment of Gowrie Colquhoun Aitchinson as manager proved to be of immense benefit — his previous experience on various railways stood him in good stead and he certainly contributed to the success of the Snowdon railway. His salary was £300 a year plus a share of the annual profit and local villagers stated that it was an impressive sight to see him in the early morning going to work from his house at Glan Padarn on his magnificent white horse!

Gowrie Colquhoun Aitchison, the railway's manager in 1896.

Derek Rogerson, manager, 1990.

On 23 May 1895 the subscription list of the company's shares was opened. The issue was of 6,343 shares at £10 each and mortgage debentures of £100 each. The list of subscribers did not indicate any interest by local people and most of the shares were bought by people from outside the locality. The first share was bought by a 'Mrs Alice May Corbett of Wrexham' — her only share! 53 persons purchased a total of 5,454 of the 6,343 shares allocated — Captain Stewart, agent to the Vaynol estate, bought 250 as did the principal directors of Holme & King (Arthur Hill Holme and Charles Wilden King). Other shareholders were Harmood Banner, the company chairman, with 400 shares, Colonel Cragg had 200, R.E.C. Naylor 400 and Clinton James Wilson Holme with 200.

It was the intention of the directors to build a two-storey

hotel on the summit, on an acre of land, with seventeen bedrooms for those who desired to sleep on Snowdon to see the sun rise. The summit at that time was occupied by a cluster of wooden huts from which two tenants, Robert Owen and Thomas J. Roberts, sold the most acceptable of refreshments. Naturally, the huts were impediments to the construction of the hotel, and in August 1895 at Porthmadog Assheton Smith applied for the possession of 'certain premises on the summit of Snowdon on which portions of the two huts or other buildings used as licensed refreshment homes are erected, which premises are let to the defendents as tenants from year to year and which tenancy was duly terminated by notice to quit expiring on the 12th day of November 1894'.

The judge gave 'judgement for plaintiff as regards ownership of land on which part of it stood according to the line of boundary agreed upon by plaintiff'.

By the summer of 1895 the huts to accommodate the working force had been erected and work progressed at a reasonable pace but it was obvious that the trains would not be running by July as had been envisaged. However, by the middle of June most of the track had been prepared but there was an appreciable amount of work to be done on the seven bridges, the cuttings and the two viaducts — most of the work had to be done by the use of spade, pick and by pushing wheelbarrows.

On 28 June 1895 the first loco arrived in Llanberis from Winterthur in Switzerland — a historical day indeed! The loco had been conveyed by rail to Antwerp, then by sea to London docks and then carried on an LNWR flat truck to the mainline station at Llanberis. Temporary rails had been laid on the roadway from the LNWR terminus to the mountain railway station and as the loco was pulled by a team of horses, the rail on which the loco had travelled was then lifted and re-laid on the road. Slowly but surely the loco arrived to great expectation at the Snowdon mountain railway station.

The name of the locomotive was rather unusual — *Ladas* — but it was to honour the wife of the squire and landowner of the Vaynol estate, Laura Alice Duff Assheton Smith, the initial letters of whose name spell out the word 'Ladas'. The invoice in respect of the purchase of the loco was surprisingly made out to 'Holme & King', the contractors, and not to the new railway company.

le 10th June, 1895

Messrs Holme & King, Liverpool Winterhur

Societe Suisse pour Construction De Locomotives et De Machines.

Pour les articles suivant détails expédiés pour votre compte et à vos risques et perils par Chemin de fer a l'adresse de Messers John P. Best & Co. a Anvers payables suivant contrate.

1 Rack-Rail Locomotive 'LADAS' for Snowdon Railway
Engine No. 1
Boiler No. 1544
Fabrie No. 923

complete according to specification including Royalty for Abt system and automatic steam brake delivered free Antwerp £1525.

Even today you will find that employees of the mountain railway will always refer to the engines by their number and it is very rare indeed that they quote their names. It is invariably the passengers that refer to the locos by their names.

During 1895 the main work of constructing the two viaducts by Chambers of Manchester was proceeding with some difficulty. The lower viaduct was 166 yards long with 14 arches, each of 30 feet, and the lower arch over the road was to be 38 feet, 4 inches. Since the gradient was 1 in 8, each arch was four feet higher then the lower one. The upper and smaller viaduct was 63 yards in length with four arches. Scaffolding for building the two viaducts was readily available — the estate had given permission for the contractors to cut all the necessary trees that grew on the nearby mountain slopes. Some of the letters that were exchanged between the estate and the contractors referred to the difficulties encountered.

> Dear Capt Stewart,
>
> There is another lump of rock to the quarry which I should be much obliged for your permission. It is situated about 50 yards from the lower viaduct at the bottom of the hill. It has been quarried before and being very good stone, with natural flat and square beds, will expedite the building contractors.
>
> Yours truly,
> Clinton Holme.

Mr Oswell, the Engineer, had also informed the correspondent of the local newspaper that 'we are giving level crossings and cattle creeps to whoever tenant farmer asks for them and we have 1½ miles of walling to do on each side of the line'.

The rail system was to be the Abt system so successful on the Swiss mountain railways and based on the fact that the distance between each and every tooth was the same so that there was always a connection between the pinion and the rack below. The work of supplying the trackwork (including the rails and the racks) was allocated to Richard Cammell & Co. o ` Sheffield. Since the track had to be laid on steep slopes it was necessary to erect steel pegs in concrete to hold the sleepers so that the track would not slide off.

When we realise that the workforce did not have at its disposal such 'convenient' machinery as the helicopter and JCB of today, it was indeed an engineering feat that the whole track — nearly five miles in length — was laid between Llanberis and the summit of Snowdon in the amazingly short period of 72 working days. On average, some 1,200 yards were laid every day, and indeed during one extraordinary effort it was recorded that 1,500 yards was laid on what was very rough and stony ground. The machinery was brought up the mountain on horse-drawn sledges until this laborious task was replaced by using the locomotive that had arrived from Switzerland.

As mentioned, the men were accommodated in special wooden huts erected specially for them. It is rather surprising to realise that no work was carried out on week-ends, especially in view of the fact that originally only nine months had been agreed on in the contract for the whole project.

There was without doubt some disagreement as to the ownership of the land on the summit itself. Sir Edward Watkin and Mr Assheton Smith did not see eye to eye as to the actual boundaries and furthermore Assheton Smith felt it necessary to bring a court case against Owen Roberts, the tenant of one hut on the summit. On 13 August 1895 at Porthmadog, judgement was given in favour of Assheton Smith and as owner of the land in question he agreed that Owen Roberts could remain as tenant of the hut until the end of the season.

At the end of August 1895, again at Porthmadog, one Robert Owen applied for renewal of his licence to sell drinks at his hut. Assheton Smith opposed the application, but Robert Owen in his defence testified that he had sold refreshments on the summit for thirty years and no one had complained. He had kept his property in good condition and had indeed erected another safety wall to protect his property. In addition he had built a new hut with four bedrooms and the whole work had cost him five hundred pounds.

Horatio Lloyd, the solicitor who appeared on behalf of the railway company, applied for a licence for the hotel that was to be built on the summit at a cost of £500. The inhabitants of Beddgelert, the village on the other side of Snowdon, were vehemently opposed to the erection of a summit hotel and the petition presented on their behalf 'outlined the excessive number of buildings on such a small piece of land'. There was no need for a hotel and furthermore they enquired as to the outcome of any 'misdemeanour and misbehaviour by drunks in a building so far removed from police supervision'.

Again Assheton Smith was unsuccessful. The justices at

Porthmadog refused the application for a licence to sell alcohol at the proposed summit hotel.

Meanwhile the Victoria Hotel at Llanberis was getting spruced up for the influx of visitors:

> A hotel with a large coffee room, bathroom, billiard room, smoking room, and upwards of 40 bedrooms with beautiful laid-out grounds. Ponies and guides to Snowdon and boats on the lakes!

Work on the railway proceeded and by September 1895 one and a half miles of rail had been laid and the lower viaduct had also been completed by early autumn. But what of the progress of the railway in the adjacent valley, 'The North Wales Narrow Gauge Railway'? The track had been laid as far as the village of Rhyd-ddu with hundreds of passengers travelling on the train as far as the so-called 'Snowdon Station' — as the village station had by then been re-named. Unfortunately, the passengers were unaware that there was a further three miles of quite steep climbing in front of them along a quite treacherous path to the actual summit itself.

The slate quarrying industry was by then enjoying an upsurge and by the end of the year the 'Ruthin Charities' who owned a substantal amount of land in Llanberis were offering pieces of land for sale:

> Llanberis is the chief quarrying district in Wales and if there were energetic people to be found on the face of the earth, then they surely would be at Llanberis because they had moved heaven and earth to persuade English capitalists to come there and lay for them a railway to the summit of Snowdon. The result of this was that after the opening, the small town would spring up and property value would increase.

On the other side of Llyn Padarn, the railway line to convey slates from the quarry to the port at Felinheli not far from the mansion at Vaynol Park, was ready for use.

Holme & King in a progress report in November 1895 were extremely hopeful that should the weather remain favourable the railway track would shortly be ready for use. The correspondent of the *The Caernarvon & Denbigh Herald* had travelled up the mountain and witnessed on the high slopes 'men hanging on like grim death to whatever solid substance offered protection and they were blue in the face'.

The arrival of the two locomotives, *Ladas* and *Enid,* was a blessing indeed and as a result machinery could be carried up the mountain to the working areas. The shape of the locos did cause some consternation to the locals, one apt description being that 'they have the appearance of having being built purposely to plunge their heads down into the earth'. The benefit derived by using the two locos can be gauged by the fact that on one day *Enid* carried twelve tons of track steel bars and ballast. It is recorded, whether true or not is insignificant, that one Irish labourer employed on the project — no doubt concerned that the completion date had long passed by — suggested to the company officials that they 'should cut off the summit of Snowdon so as to spare the cost of laying half a mile of track'! It is worth noting that during excavation work at the summit seashells were found over three thousand feet above sea level.

Though the two locos were most welcomed, the horses were still retained and were extremely useful in carrying out remedial and difficult tasks. The destiny of two of those hard working beasts certainly aroused my interest.

Daylight — bought for £35, died of lock-jaw.
Strawberry — bought for £31, died of sunstroke.

The writer recently held one of the horseshoes that belonged to one of the shirehorses and I must confess that I had never handled a horseshoe as heavy and as big — it weighed three and a half pounds and measured seven inches across. Since the poor horse had four of these it presumably had to carry a stone in weight on its feet in addition to the ardous task of pulling the sledge and its contents as well.

The land on both sides of the track was farmed by numerous small-holders — who not only were the tenants of the Vaynol estate but also supplemented their earnings by working in the huge Dinorwig Quarry during the day. Indeed, their working day was a lengthy one, having to walk to the quarry in the early hours and returning up the steep hill back to their homesteads in the early evening. Assheton Smith had promised them that their rights would be protected and the content of a letter to Captain Stewart, the estate agent, from Colonel Clinton Holme of Holme & King in July 1895 denotes the desire of the landlord and his staff to pursue any complaints raised by the tenants.

> In answer to your letter from William Owen, which I now return, I have given strict orders about keeping the dogs tied up and should further complaints be made I will have them removed altogether. There are only two dogs belonging to the huts.
>
> With regard to the grazing, I have made enquiries and find that the horses belong to the man at the top half-way huts (Moses Williams) who does all the provision carrying and he tells me that they are turned loose on the mountain at night and that he pays the tenant for so doing.
>
> Yours truly,
> Clinton Holme.

CHAPTER SEVEN

Problems Exist

At three minutes to eleven on Thursday morning, 9 January 1896 the first train to reach the summit of Snowdon left the base station at Llanberis. This was not a passenger carrying train but conveyed a number of company officials, some workmen and representatives of the contractors on what was a test-run. One cannot but commend the unbounded enthusiasm of the directors in 'arranging' a train to go to the summit of Snowdon in January of all months! Fortunately all was in order and having reached the summit, the loco and two carriages stayed for twenty minutes before returning in triumph to Llanberis. A historical day indeed!

There still remained some work to be done on the various station buildings and some fences were to be erected. The proposed signalling system had also to be attended to. The intention of the contractors was to transfer the line and all buildings to the Snowdon Mountain Tramroad and Hotels Ltd by the re-arranged date — Easter 1896. Even in the midst of all the excitement that certainly existed there still was some opposition to the project. The intention to build a hotel on the summit was certainly not to the liking of one Harold Hughes as his letter to *The North Wales Chronicle* on 8 January 1896 proved:

> Is money everything?
>
> Is not the mountain sufficently desecrated by the Snowdon Railway? Must we for a little gold add further insult by crowning it with a structure so out of harmony with

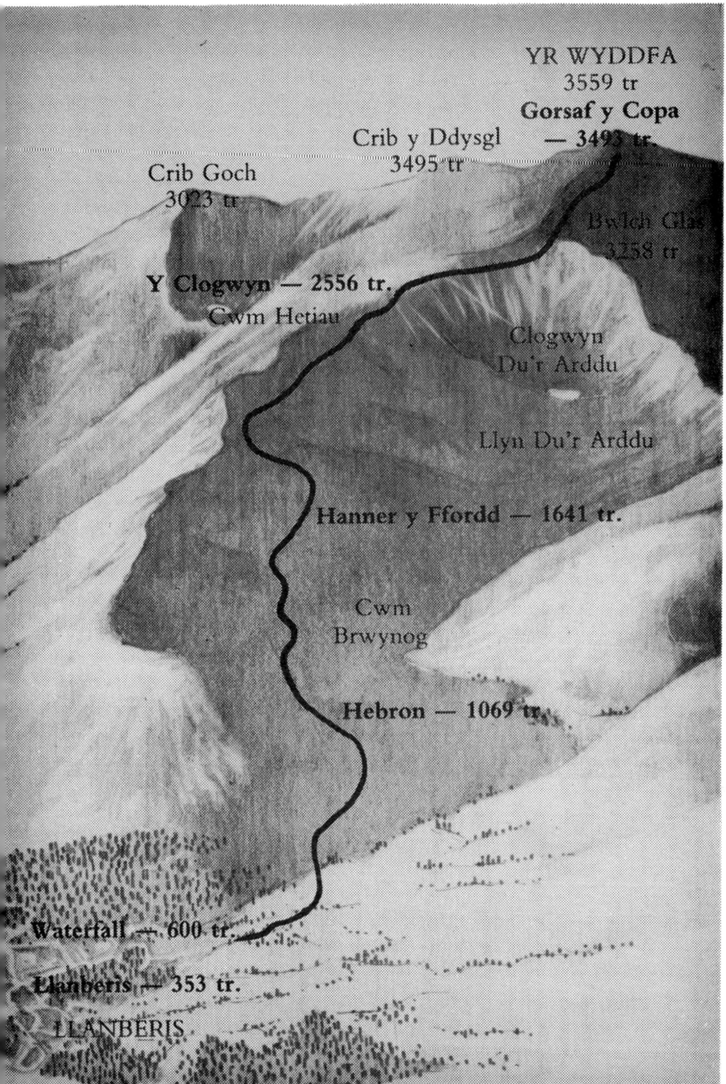

A map of the railway.

> all the surroundings as a first class hotel? A leading geologist well known beyond the borders of Wales, who has taken up residence amongst us for the purpose of our county, in a recent lecture stated that it was a disgrace to the Welsh nation to have permitted the construction of the railway on Snowdon. The huts which exist at present at any rate are more in keeping with the surroundings of the summit than a first class hotel could ever be. We feel that any true lover of the mountains will regard the idea of this first class hotel with anything but sorrow.

From time to time Assheton Smith voiced his opinion on certain proposals of the company, and not always in a favourable manner. For instance, though it was originally intended to erect two platforms at Hebron station the idea did not materialize and it was whispered that the landowner did not favour such a necessity. The original intention was to electrify the line and build a dam between Cae Newydd and Brithdir in Cwm Brwynog for that specific purpose, but the idea was eventually abandoned.

In order to maintain as much publicity as possible, various newspaper proprietors and editorial staff were invited to travel on the line and, no doubt, in due course commended their readers to travel likewise when the railway would be officially opened.

Since the venture was a private one it was not legally necessary to carry out an inspection of the line prior to an official approval. Nevertheless, Sir Francis Fox prevailed upon his personal friend, Major F.A. Marindin, Chief Inspector at the Board of Trade in London, to visit Llanberis and carry out an 'unofficial' inspection of the track and locos. The Board of Trade did not look upon Major Marindin's intention with much favour as it was felt that such a visit might give the impression that the Board was associating itself in an official capacity with a private project. The Major duly arrived at Llanberis and carried out an inspection on 27 March 1896, by having two flat carriages filled with iron bars and cement bags — 18 tons in all — and then pushed up the mountain to the top. On the return journey the Major ordered the driver to go as fast as was possible so as to check whether the automatic brake functioned at 5 m.p.h. Again, when a similar test was carried out on the lower slopes where there was a steep gradient, the independent brake system was effective in holding heavily-laden carriages.

In his report to the Assistant Secretary of the Board of Trade dated 3 April 1896, the Major remarked on the fact that station buildings had been erected at Hebron and Waterfall (which is no longer in use) but that work at Halfway and Clogwyn stations had not been completed. In addition Major Marindin stated that:

> . . . the work on the viaducts has been well constructed and they appear to be quite stable. The girders of the underbridges have sufficient strength and were satisfactorily tested. The brakes on the train are thoroughly efficient. When detached from the engine each carriage can be stopped or let down slowly as may be desired on the steepest gradients by means of the brake attached to it.
>
> There was much work to be done at the summit station. I do not think that objection need to be made to the opening of the line, subject to a substantial buffer being built at the end of the platform at the summit.
>
> At Hebron there should be a second platform provided. It could also be necessary to have some restriction on the running of the trains at certain points during gales of wind. Wind pressure should be considered dangerous when it reaches 15 lbs per square feet, i.e. a bad storm.

As a result of the Major's recommendation a machine was to be installed at Clogwyn station to measure the speed of the wind. At such a high point on the mountain the windflow is unimpeded and when low sided carriages were in use passengers invariably stood up to see the village of Nant Peris 2,000 feet below. In doing so their hats were often blown over the precipice and down the mountain slopes, where the villagers of Nant Peris would come at week-ends to obtain 'new headgear'! More than one quarryman of Nant Peris showed off his newly acquired felt hat to admiring

The summit of Snowdon. Note the very open situation of the huts and that the signals had been installed when the photograph was taken.
(Photograph by courtesy of Iwan Roberts.)

The station at Llanberis and loco' Wyddfa *ready to start the ascent.*
(Photograph by courtesy of Iwan Roberts.)

colleagues on Monday mornings, and indeed to the locals the area became known as 'Cwm Hetia' (Hat Valley).

It was eventually agreed between the directors and the contractors that the railway would open to the public on Easter Monday, 6 April 1896 and a very energetic publicity campaign informed all and sundry of this intention. By now the venture had cost £76,000, part of which had been spent on refurbishing the Royal Victoria Hotel and purchasing land at Tŷ Clwb to build the base station and various offices and stores.

Though a slight accident had occured on Easter Saturday morning, 4 April, when one carriage had come off the rail, this did not deter the company from proceeding with the avowed intention to officially open the line within two days.

On Llanberis lake during the great frost, 1895-6.

CHAPTER EIGHT

The Official Opening

Easter Monday was a sunny spring day with blue skies hovering above the remains of the wintry snow amongst the crags and cliffs of Snowdon.

At daylight a number of workers were ordered to walk the whole length of the line and to inspect in detail every inch of the track and in addition a 'pilot train' followed them in the early morning. Everything was found to be in order.

Though Easter Monday was an extremely historical and important day to the inhabitants of the area, there were other distractions as well. A total of 5,179 day return tickets were sold at Bangor railway station and over 400 people went on the pleasure steamer *Torbay* that sailed between Bangor and Beaumaris in Anglesey. At Llandudno the final of the Welsh soccer cup between Wrexham and Bangor drew a gate of 10,000! At the Oval ground in Caernarfon, horseracing was the attraction and though Captain Stewart was one of the directors of the Snowdon Mountain Tramroad & Hotels Co. he was one of the judges at the Oval races rather than being an official passenger on the first official train up Snowdon.

The directors of the new company had agreed that an opening ceremony was not necessary, though naturally a large crowd had gathered at the station to see the first train and its two carriages pull in by the new platform to take eighty or so passengers to the summit. (Though there are no records to state which loco, it does seem that the honour of

After crossing the bridge over the Hwch river, the train climbs to Hebron station.
(Photograph by courtesy of Iwan Roberts.)

being the first locomotive to carry passengers to the summit fell to *Enid.*)

It had been arranged that the train should leave the base station at half past ten in the morning but since the special main-line train with its hundred passengers from Chester and the North Wales Coast had not arrived at the local LNWR station the departure of the Snowdon train was delayed until ten minutes to eleven. As *Enid* passed the Waterfall station the other loco *(Ladas),* as agreed, moved on to the side platform to take the rest of the passengers. The driver was William Pickles, an experienced driver from York, with his nephew of eighteen years of age as his fireman.

Within the hour both trains reached the summit, but because of the low cloud base it was agreed that *Ladas* and its two carriages should start the return journey at half past twelve and it was further agreed that when it reached Clogwyn station a bell should ring giving a signal at the summit that *Enid,* driven by John Sellers, should start its journey back to Llanberis. When *Ladas* was some hundred yards from the bridge that crosses the public footpath above Clogwyn station (and indeed on the steepest gradient on the line) the engine lost connection with the rack and as a result it hurtled down the track at an uncontrollable speed. The automatic brake was not effective and when the loco was at the bend on the track above Cwm Glas, *Ladas* came off the track and plunged over the cliff into the depths of Cwm Glas below. Fortunately William Pickles and his young nephew had been able to jump from the footplate and suffered no injuries. Gowrie Aitchinson, the manager, who was travelling in the first carriage behind *Ladas* (carriages being uncoupled as was and still is the custom), applied the brakes and was able to stop the carriage within 30 yards and so prevent a terrible tragedy. It was rumoured that a certain Scottish gentleman called McFarlane was climbing the high slopes above Cwm Glas at the time and when he saw a huge locomotive 'flying in mid-air' he was so petrified that he immediately packed his bags and returned forthwith to Glasgow!

As *Ladas* came off the track it damaged the telegraph poles and cables at the track-side with the result that a 'rogue' connection rang the summit bell. John Sellers naturally took it as a signal for him to start back with *Enid.* By now the mist had become quite thick so Sellers took extra care.

Mr Aitchison suddenly remembered that *Enid* had yet to come down so he ordered the guard to run back up the track to warn Sellers of the accident lower down the line. Before the carriages had been brought to a halt by the application of the brake by Aitchison, he had told the worried passengers in the *Ladas* carriages to sit down in their seats, but unfortunately one passenger, Ellis Roberts, the proprietor of the Padarn Villa hotel at Llanberis, had witnessed the young fireman jumping off the footplate and in his panic he did likewise and in his fall suffered terrible injuries.

The remains of L.A.D.A.S. *in Cwm Glas following the accident of 6 April 1896.*
(Photograph by courtesy of the Gwynedd Archives Service.)

The scene following the accident, 6 April 1896. L.A.D.A.S. *plunged over the cliff at the point where the figure stands. Two carriages are seen in the distance lying on their sides.*
(Photograph by courtesy of the Gwynedd Archives Service.)

Loco' 'Enid' following its collision with L.A.D.A.S.'s *carriages on 6 April 1896.*
(Photograph by courtesy of the Gwynedd Archives Service.)

Two of L.A.D.A.S.'s *carriages near the Clogwyn station after colliding with* Enid.
(Photograph by courtesy of the Gwynedd Archives Service.)

Dr R.H. Mills Roberts,
the doctor of Dinorwig Quarry
at the time.
(Photograph by courtesy of Emyr Jones.)

The grave of Ellis Roberts
in Nant Peris cemetery.

Unfortunately John Sellers did not have any warning about the fate of *Ladas* and the two carriages. As *Enid* came on to the spot where the carriages were, the men who were carrying poor Ellis Roberts 'had considerable difficulty in getting out of its way'. Enid crashed into the carriages which in the impact were hurtled forward along the line, but were eventually turned into the passing loop at Clogwyn station. The passengers had to walk down to Llanberis and no further trains were run that day or subsequent days.

Ellis Roberts who had severe injuries to his leg and thigh was carried down to Clogwyn station where railway staff trained on first-aid were able to apply a tourniquet to stop the terrible bleeding. They also filled empty beer bottles with hot water which were used to try and keep the injured man as warm as possible until a doctor arrived. The accident occurred about a quarter to one in the afternoon.

The remains of L.A.D.A.S.'s *boiler in Cwm Glas after the accident. It was later sold to Dinorwig Quarry for £190.*
(Photograph by courtesy of the Gwynedd Archives Service.)

A message was taken to Dr R.H. Mills Roberts, the doctor at the Dinorwig Quarry hospital, and he immediately arranged to travel up to Clogwyn station. He was informed by the official at the base station that it would not be practicable for him to be conveyed by train up to the accident spot as it would take two hours to get enough steam, so Dr Mills Roberts was able to use a pony for part of the way, the remainder he walked and arrived at Clogwyn about three o'clock. Immediately on arrival he attended to Roberts's injuries and, using a carriage door as a temporary stretcher, volunteers carried him down to Llanberis where they arrived about 6 p.m. After consulting with the local doctor, Dr Lloyd Williams, it was agreed to amputate Ellis Roberts's leg. Sadly, he died at midnight from his injuries, having lost a great deal of blood.

On the following Wednesday, 8 April, Major Marindin having travelled from London accompanied Sir Douglas

Fox, Colonel Holme and other company officials to the location of the accident. The Major then descended down the slopes of Cwm Glas to inspect the remains of *Ladas*. The second loco involved in the accident, *Enid*, had been de-railed and of necessity had to be left on the mountain side above Llyn Du'r Arddu.

The following comment written by Gowrie Aitchison in his record book certainly took my eye — 'ENID was considerably injured in an accident April 6 1896'.

CHAPTER NINE

The Inquest

At the adjourned inquest the Coroner, Mr J.H. Bodvel Roberts stated that he had been informed by the Board of Trade that the Snowdon railway was not authorized by any Act of Parliament. It was a private undertaking and as such Parliament could not interfere.

One witness, S.W. Williams, stated that about a quarter of a mile from the summit on the return journey he had heard 'a terrible noise' and had noted with much concern 'that the train had gathered great speed'. He had seen Ellis Roberts, who was sitting by the door, jump out of the carriage. Another man also jumped out (later ascertained to be a Mr Jackson from Oswestry) and had suffered some abrasions to his head. When the carriage was eventually stopped he, like other passengers, had gone to give assistance to Ellis Roberts who was still conscious at the track-side.

Other witnesses estimated the speed of the train to be between 3 and 12 m.p.h. Gowrie Colquhoun Aitchison, the general manager, stated that Major Marindin had inspected the line on 27 March. A special train had travelled to the summit on the morning of the accident and all was well. He agreed that a few minutes before the accident the loco had gathered speed and that the automatic brake had come on. When it was re-set he ordered the driver to proceed slowly and with care. But there was a significant increase in the speed and realising that some passengers were getting rather worried he had ordered them to sit down and not to panic. When he applied the brakes the loco stopped

The train about to depart from Llanberis station at the turn of the century. The man standing on the right is believed to be Mr Aitchison. There were no glass windows in the carriages at this time.
(Photograph by courtesy of the Gwynedd Archives Service.)

within 30 yards. Though the carriage door was locked it was possible for any determined passenger to lean out of the window and open the door. He had requested the guard to run back up the line to warn the driver of *Enid* of the accident and he had also requested another person to run down to Clogwyn station to request medical assistance from Llanberis. Unfortunately the guard did not warn John Sellers, the driver of *Enid*.

Amadeus Taechella, an engineer from the Abt company in Switzerland, gave his evidence in French. He informed the inquest that he had come to Llanberis in July 1895 to train personnel to drive the new locos. He had remained for five weeks and had travelled up and down the mountain regularly and during the period did not see anything that was not in order.

When William Pickles, the *Ladas* driver, gave his evidence he stated that he was an engine driver of 20 years experience and had been at Llanberis since July. He had driven *Ladas* up and down the mountain, five or six times daily, carrying equipment and machinery up to the summit. On Easter Monday to loco had been thoroughly inspected before departure and during the ascent everything was perfect but a few yards after leaving the summit the loco seemed to pull to the left. Then he heard a screeching noise and the train was by then travelling at an excessive speed. Immediately he ordered his young fireman to jump out. He himself remained in the loco and tried to apply all the brakes but to no avail. Eventually he had to jump out himself when the train had reached a speed of nearly 20 m.p.h. He estimated that *Ladas* had travelled some 80 yards out of control before he jumped from the foot-plate.

John Sellers, the driver of *Enid*, remarked that soon after leaving the summit he felt a slight irregular movement on the track and immediately applied the brakes. *Enid* could, in his opinion, have suffered the same fate as *Ladas* but for the collision. Fortunately none of the passengers were hurt.

In his evidence Sir Douglas Fox, the Advisory Engineer,

The hotel which had been proposed for the summit of Snowdon.

(Photograph by courtesy of Reg Hughes.)

The staff of the railway in 1896. Seated in the chair is Gowrie Aitchison, the manager. Others in the photograph are William Pickles, his nephew Billy (back row, left hand side), John Sellers (back row, right hand side), Tom Williams (driver), Charles Parry (driver), Hwfa Williams, Tom Williams, Robert Lewis, John Llewelyn Williams, Owen Griffiths, John Morgan, John Hayes, John Hughes, Humphrey Williams and David Lewis.
(Photograph by courtesy of Mrs Enid Foulkes.)

stated that he did not wish to speculate but it was not difficult to realise that there was a relationship between a landslip and the accident, which in his opinion occurred when the loco was near the cutting in the rock-face and not lower down as some believed. At that point the inner part of the track was supposed to be half an inch lower than the outside part but it seemed that the difference on the day was two and a half inches and such a discrepancy was clearly visible in the landslip which could have happened before the train came onto the track. Sir Douglas believed that the snow had gathered in the cutting and that ice had formed under the rails. When the sun came out later in the morning the snow and the ice melted and as a result the foundations had been moved. It should be remembered that such a foundation was not of hard rock but of stone and hard ground.

Dr Mills Roberts, the Dinorwig Quarry doctor, in evidence stated that he had been informed of the accident at twenty minutes to two in the afternoon. When he arrived at Clogwyn station at 3 o'clock it was apparent that Ellis Roberts had been seriously injured and had lost a lot of blood. Ellis Roberts was brought down to the Padarn Villa hotel at six o'clock — five hours after the accident!

Around midnight he had consulted Dr Lloyd Williams and it had been agreed that it was necessary to amputate poor Ellis Roberts's leg but unfortunately he died in the early hours. In his opinion it seemed that the deceased had hit his head against the rockface when he jumped out and that he had rolled back on to the track and was hit by the carriage step.

The Coroner, Mr Bodvel Roberts, stated that Ellis Roberts could well have survived the accident had he remained seated in the carriage like the remainder of the passengers.

The foreman of the jury gave the verdict '. . . that the deceased, after the loco came off the rack and having seen the driver and fireman jump off the machine when it was moving at an excessive speed jumped from the carriage and met his death. There is no evidence to show why the machine left the rack' (*Herald Cymraeg*, 20 April 1896).

The summit station in the early days, the huts standing above it on the slopes.
(Photograph by courtesy of Iwan Roberts.)

CHAPTER TEN

Back in Business

During a conversation between Gowrie Aitchinson and a reporter of *The Caernarvon & Denbigh Herald* at the end of April 1896, Aitchinson referred to a meeting that had been held in London to discuss the accident and mentioned some improvements to the track that would prevent further such accidents. He confirmed that repairs had been carried out and indeed one train had been able to go the summit since the accident. No decision had been made regarding the re-opening of the line.

But in Liverpool, one Customs and Excise officer by the name of John Ord believed that in the course of his daily duties he had stumbled across the cause of the accident. In a letter addressed to the Chairman of the Board of Trade on 10 April 1896 he boasted that a box containing twenty four mechanical parts had arrived at Liverpool Docks that morning. In his opinion they were urgently required by the Snowdon railway since they seemed to be brake parts. He wrongly assumed that the brake parts had thus not been fixed to the locomotive *Ladas* on the day of the accident as they should have been and he felt that an official of the Snowdon Railway Company who had come to Liverpool to collect the parts not only seemed in a great hurry to obtain them but also behaved in a rather dubious manner. Ord, because of his doubt, had secretly marked each part so that if they were fixed back on the loco before the inspection they would be found out. He earnestly believed that there was a conspiracy afoot and told Major Marindin accordingly.

A respite at the Half Way station. Note the open carriage. It is no wonder that so many hats were blown over the cliff, and that the people of Nant Peris dubbed the area 'Cwm Hetia' ('Hats Valley').
(Photograph by courtesy of Ken Jones.)

When the Major discussed the matter with officials of the Snowdon Mountain Tramroad & Hotels Co he soon found that all was in order. He was informed that the brake parts certainly were urgently required for another loco at Llanberis: since *Enid* had been damaged and *Ladas* destroyed there was an urgent need for the reserve engine to be prepared for use as soon as possible. The Winterthur works had promised early despatch and that was the reason for sending an official all the way to Liverpool to fetch the parts. It is not known what became of John Ord at Nelson Docks after his vain attempt to dishonour the Snowdon Mountain Railway.

Many questions had been asked in Parliament about the accident. On 25 June 1896 Massey Mainwaring requested further details but Mr Ritchie, the Minister responsible, in reply stated that the opinion of the inspector was that a landslip seemed to be the cause of the accident and that the company did not intend to carry any passengers until all precautions had been taken to prevent a recurrence of the accident.

Throughout the immediate post-accident period the directors were adamant in their intention to continue with the project. In May 1896 an advert in *The Caernarvon & Denbigh Herald* invited tenders from contractors to build a new hotel at the summit where the cabin tenants still carried on with their trade, as the following memo from the Baron Hill Estate office reveals:

> Mr Assheton Smith will agree to let Mr Thomas Roberts continue in occupation of his portion of the hut on the summit of Snowdon for this year 1896 at the rent of ONE POUND. The boundary along the watershed between Vaynol and Baron Hill to be defined by iron pegs.
>
> Mr Thomas Roberts agrees to this arrangement.
>
> Witnessed by Leo Hepworth, Baron Hill Estate Office, Beaumaris.

Naturally the cause of the accident became a topic of conversation amongst experienced engineers. But the Abt company was not willing to accept any allegations that a mechanical defect was the cause and was emphatic that no alterations were necessary to the locos nor to the track either, though the Board of Trade had suggested fixing guard rails at the side of the track.

Nevertheless a representative of the Abt company in the United kingdom, Frank Passmore, on 19 March 1896 (i.e. before the accident) carried out a track inspection and had submitted a critical report on its condition.

In his opinion there were at least four areas requiring immediate attention:

1) The ends of the sleepers were protruding beyond the ballast in many places and some sleepers were loose.
2) Stones had been placed on top of the sleepers with no ballast in the centre.
3) Some of the rails had been incorrectly fixed with some variation in the space between the rails.
4) No grease had been applied to the rack nor to the pinion.

He suggested that kerosene and oil should be mixed so as to prevent the oil from freezing on the higher slopes. Furthermore, he felt that three quarters of the line should be re-set as the rack teeth had been broken in places.

His report was solely for the attention of the Abt company and it is not known whether the Snowdon railway company received a copy or acted on the recommendations. After the accident the relationship between the company and the contractors was not ideal and as a result the company employees had become responsible for the maintenance of the line — the directors had purchased all the remaining stocks of rack from Cammell and Co, and even to this day steel bars bearing the name 'Cammell 1896' can still be found on parts of the track.

In May of that year the fixing of safety rails became a priority with the hope that it would be possible to re-start passenger services at least as far as Hebron station before July. Unfortunately this was not to be. As a result of the advertisement in the local *Herald Cymraeg*, some tenders to build the hotel had been received but all were far too high for the directors to consider. There was some ill-feeling between the directors and the contracors regarding the standard of work on the bridge over the footpath by Clogwyn station and also some fencing and drainage work did not meet with their approval. Indeed matters deteriorated to such an extent that Sir Douglas Fox deemed it necessary to warn Holme & King that all work had to be completed by 20 June.

Assheton Smith throughout this period had insisted that the boundary walls between his land and the track had to be of stone but realising the financial implications of the accident, he agreed that fencing would suffice, and probably this is the reason why there are no stone walls to be found after the first mile or so of track.

The extra financial burden arising out of the accident — £9,000 — prompted Sir Douglas Fox to suggest that not only should safety rails be fixed on both sides of the track but that a connecting bar should also be fixed beneath the loco and the track. The Abt company did not take too kindly to this suggestion as it might cast doubts on their engines. However, the directors agreed with Sir Douglas and a local firm, De Winton of Caernarfon, was given the contract to produce the safety rails. In July 1896 the line had still not been re-opened.

Climbers and hill walkers still went to the top of Snowdon on foot and one or two, as the following verse denotes, slept in the huts at the summit:

> Some folks I see with great surprise
> Get up to see the sun arise,
> But tranquil I remain in bed
> The sun sees me, get up instead.
>
> (Visitors' Book, 1896)

At the end of July 1896 the contractors Holme & King left Llanberis after two years of dissapointments, frustrations but much satisfaction as well. One local 'poet' recorded the departure in verse — a loose translation from the Welsh would be:

When the rail indeed was finished
And the 'navvies' done their stint,
All the 'pigs' went over Menai
All the 'snails' went back to Flint;
All the Irish back to Ireland
All the English — every man,
But remaining in Llanberis
Children, aye, to more than one!

The 'pigs' and 'snails' being the expressions used when jocularly referring to the inhabitants of Anglesey and Flint.

On 20 August 1896 the fourth engine, *Snowdon*, arrived at Llanberis, a much more powerful loco than the others. Some trains were being run as far as the Waterfall station on a fairly regular basis by September — a month of wind and rain so it seems. A thunderstorm on the ninth day of the month had destroyed Owen Roberts's hut on the summit; the inhabitants dashing to the small cafe where there was a lightning conductor!

The winter of 1896-97 was very harsh, the base station at Llanberis in January was under eleven feet of snow and indeed in early April the cuttings on the higher slopes had six feet of hard snow and ice. But on Easter Monday 1897 — twelve months since the accident — a special train left Llanberis in the early morning and travelled as far as Clogwyn. By 10 o'clock it had returned and the officials were satisfied that the track was in order. A large crowd had assembled to celebrate the re-opening of the railway. Thirty four passengers with Gowrie Aitchison and Major Turner went up on the first official train and four more trains followed during the day. The safety rails had been installed on the track as far as Clogwyn station and no train proceeded beyond that point. At five o'clock the last train safely returned to Llanberis to be received with a mixture of enthusiasm and relief.

The Snowdon Tramroad and Hotels Company was back in business!

There were one or two claims for compensation resulting from the accident – a Miss Kilshaw from Liverpool was one successful claimant, though no official decision had been made as to who or what was responsible for the accident.

On the other side of Snowdon the North Wales Narrow Gauge Railway, later to become the Welsh Highland Railway, was very conscious of the competition that would now exist and indeed had offered Gowrie Aitchison the post of Manager and Secretary to the company. The departure of such an experienced and dedicated official would certainly have been felt by the Snowdon railway so it was agreed that he could accept the new post with the NWNGR on the condition that such work would not interfere with his duties at Llanberis. A surprising decision, maybe, in view of the competition that existed.

Moel Siabod, another engine, duly arrived at Llanberis at the end of 1897 – there were four in use by then. It was also possible to announce that the total cost of the whole project amounted to £76,152 – twelve thousand pounds more than the estimate.

Regrettably it is not possible to obtain photographs of those early years in the history of the railway. One reason could be that the company, for reasons unknown, would not let any unofficial cameraman take photographs. They had allocated the task to a certain L.C. Symonds from Llanberis whose photographs of the period seem unobtainable today.

As the popularity of the railway increased it was necessary for the village hotels to receive as much publicity as possible and in 1897 and 1898 potential customers were reminded that 'the Royal Hotel in connection with the Snowdon Mountain Tramroad, the lower terminus of which adjoins the hotel grounds' was a most pleasant establishment to stay at. The Royal Goat Hotel at Beddgelert on the other hand duly proclaimed that 'coaches meet all trains between this hotel and Snowdon station', the North Wales Narrow Gauge

Moel Siabod, *No. 5, waiting at Clogwyn.*
(Photograph by courtesy of Iwan Roberts.)

Railway terminus. By 1897 it was obvious that Llanberis had, by the completion of the mountain railway, established itself as the principal tourist and mountaineering centre of Snowdonia — a fact much acclaimed by the local people but it could easily have been otherwise.

Ellis Roberts, the only fatal victim of the accident, was the proprietor of the Padarn Villa hotel and as a result of his death the railway company purchased the hotel. Dr Mills Roberts's services were also confirmed on his appointment as the official doctor to the company with a proviso that he receive the sum of eight shillings for every case he treated! Incidentally, Dr Mills Roberts — or Dr Mills as he was familiarly known in the locality — had been a Welsh international goalkeeper and was a member of the famous Preston North End soccer team that beat West Bromwich Albion in the 1888 Cup Final. Indeed, not a single goal was scored against him in any of the cup ties of that season — a remarkable record.

1897 was an uneventful year in comparison to the previous one and it seemed that the accident had not dampened the enthusiasm of potential passengers. Indeed, during the year 12,000 went up by train, a fact that was a source of much encouragement to the directors and staff.

By 1898 the so-called 'hotel' was opened at the summit — the cost being £700, most of it built by the company's employees. This replaced the number of huts that had been there for years — John Roberts's hut having been there since 1840 according to some records.

Additional carriages were ordered from the Lancaster Railway Carriage and Wagon Co. These in the main were open carriages and during the trip up to the top it was usual for the engine to push two of these. (It is believed that the present method of using only one carriage was started in 1923.)

On 3 May 1902 the seventh annual meeting of the company was held at Llanberis and the financial report stated thet 'the line was in a thoroughly efficient state and

Llanberis station in 1925, and two charabancs owned by Creams, Llandudno, awaiting travellers.

the year showed increased profits and an improved financial position'. On 22 November 1904 the man who had exerted the greatest influence on the whole project died. He of course was George William Duff Assheton Smith, the landowner and squire of the vast Vaynol estate. New directors replaced some of the pioneers and they now directed the future of the Snowdon Mountain Tramroad and Hotels Company.

From the beginning of the century to the First World War the company went from strength to strength and a scrutiny of the accounts reveal that they showed a profit for every year, more or less. In September 1906 the local *Herald Cymraeg* newspaper revealed that 'one day during the summer of 1906 the company had received the colossal sum of £150'.

By 1912 the open carriages were in constant use though there were some disadvantages. There was a reference in one newspaper to a husband and wife who claimed re-imbursement because a piece of burning coal had been blown by the wind from the engine chimney on to their seat causing damage amounting to £2 15s 0d.

By 1912 Charles Wilden King and Clinton James Wilson Holme, the principals of the contractors, Holme & King, who had left Llanberis in rather an unpleasant manner some time before, were welcomed back into the fold and became directors. Gowrie Aitchison joined the Army in 1913 and so left the company he had served so admirably. The chairman was still John Sutherland Harmood Banner whose guidance and expertise had been such a blessing to the company in those difficult years.

By 1921 it was agreed that further locomotives were necessary to complement the four in use, namely *Enid*, *Wyddfa*, *Snowdon*, and *Moel Siabod*, and accordingly an order was sent to the Schweierische Locomotiv und Maschinenfabrik in Winterthur, Switzerland, for a further three new locos. The first of these arrived in 1922 and it was named *Sir Harmood* in appreciation of the services given by

The engine shed at Llanberis as it is today.

TO THE CONDUCTOR OF THE

No. 950 DOWN TRAIN No. 91

YOU ARE AUTHORISED TO PROCEED

FROM CLOGWYN TO HALFWAY

SNOWDON MOUNTAIN RAILWAY

TIME 8.38

DATE 7/5/89.

SIGNED

BLOCKMAN AT CLOGWYN

THIS TICKET MUST BE HANDED TO THE BLOCKMAN AT HALFWAY

A ticket authorizing travel between Clogwyn and Half Way stations.

the company chairman. (In 1928, when the company changed its name, this loco's name was also altered to *Padarn*.) In 1923 the seventh engine arrived and was called *Aylwin*. (This was subsequently changed as well, in 1978, to *Ralph Sadler* to honour the person who had been the consulting engineer from 1964 to 1977.) The last of the eight engines, *Eryri*, came to Llanberis in 1923.

Whilst I was walking along the track a few summers ago I noticed a young ginger-haired lad with a brush and a tin greasing the rack-teeth and I thought of the suggestion made by Gowrie Aitchison many years ago that a hole should be cut in the floor of the locomotive and some piece of machinery inserted so that the track below could be automatically greased. His suggestion though not accepted at the time might yet be a practical solution to an important aspect of the daily maintenance of the five miles of track.

On 18 April 1928 the title of the company was changed from the Snowdon Mountain Tramroad & Hotels Company Ltd to the Snowdon Mountain Railway Company Ltd and this is the name by which it is known at present.

Hundreds of thousands of people have been carried on the trains during the ninety years and it is with much pride that the company can boast that the only serious accident that happened over the years was the one on Easter Monday 1896.

The success of the railway can be gauged by the 120,826 passengers who were carried to the summit in 1989, by the seven vintage locos and the recently acquired two modern diesels — a total of 2,243 trips from mid March to October.

The Rolling Stock

THE LOCOMOTIVES

No 1. LADAS 1895 (No 923 in Winterthur)

The engine was named LADAS after the wife of the landowner, G.W.D. Assheton Smith, and like all the other engines LADAS was built at Winterthur in Switzerland. According to an invoice dated June 1895 the engine cost £1,525 and arrived at Llanberis via Antwerp and London. It was destroyed on its first official journey on the opening day, 6 April 1896.

No 2. ENID 1895 (No 924 in Winterthur)

This loco was named in honour of Enid, daughter of the Assheton Smiths of the Vaynol Estate. She cut the first sod of the scheme in December 1894 in the absence of her mother. During the early days ENID carried out valuable work on the transporting of machinery and truck-rails up and down the slopes when the railway was being built. The engine was also involved in the accident on Easter Monday 1896. It is the oldest engine in present use by the mountain railway and can be seen every day on the slopes. It cost £1,525 and came to Llanberis in April 1895.

No 3. WYDDFA 1895 (No 925 in Winterthur)

This loco arrived at the Snowdon mountain railway on 7 December 1895 and is of the same type of engine as LADAS and ENID. After the accident WYDDFA was the only engine in use for some months until the arrival of SNOWDON. *Wyddfa* is the Welsh word for Snowdon.

No 4. SNOWDON 1896 (No 988 in Winterthur)

SNOWDON came to Llanberis on 23 August 1896 and is of a similar shape to the earlier types. After the early years it was taken out of service and it seemed that her contribution to the mountain railway was at an end. In 1961, however, it was re- built at the Hunslet works in Leeds and within two years was on service for the first time since 1939.

No 5. MOEL SIABOD 1896 (No 989 in Winterthur)

Named after one of the mountains of Snowdonia. Arrived at Llanberis at the end of 1897. Of the same shape as SNOWDON.

No 6. PADARN 1922 (No 2823 in Winterthur)

After a lapse of 26 years without purchasing a new engine, PADARN was the first of a new type of loco to arrive at the mountain railway in 1922. The original was named SIR HARMOOD in appreciation of Sir John Sutherland Harmood Banner — the first chairman of the Snowdon Tramroad and Hotels Co Ltd. In 1928 the name was changed to PADARN.

No 7. RALPH 1923 (No 2869 in Winterthur)

This engine is of the same type as PADARN but with a larger water tank to allow it, should the need arise, to travel up and down without stoping to take on water at the Halfway station. It arrived at Llanberis during the spring of 1923. Of all the locos this is the one that has had its name changed most. Originally called AYLWIN it was altered to RALPH SADLER after the company engineer from 1964 to 1977. Recently it was renamed — just RALPH!

No 8. ERYRI 1923 (No 2870 in Winterthur)

The last of the steam engines to be purchased — ERYRI being the Welsh word for the abode of the eagles.

The shed offering light refreshment on the path up Snowdon which stands below the Half Way station. The shed does not belong to the railway company.

Railway workers being carried for the very first time in 1896.
(Photograph by courtesy Mrs Annie Jones.)

All the locomotives were bought with a guarantee that they were able to push two carriages and indeed they were so used for many years. The engines arrived at Llanberis by rail and were off-loaded at the LNWR station and then hauled to the mountain railway base by horses along a portable track laid on the main road. In due course the boiler and other parts were assembled at the station works. At first they did not have a speedometer and as a result it was more or less left to the whim of each driver to ascertain its speed. Gowrie Aitchison rectified this and had indicators fitted to each loco.

Engines number 6, 7 and 8 cost £3,649 — twice the cost of the original ones.

During the hour long journey to the summit the locos use on average seven hundredweight of South Wales coal. Naturally, since climbing is such 'thirsty' work, some hundred gallons of water are required — the water tanks of the earlier engines hold 374 gallons and take on extra supply at the halfway stage. The tanks of the latter ones hold more.

There are three brakes on each, a handbrake on each driving axle, another brake operated by steam, and the automatic brake which operates when the engine is travelling at 5 m.p.h. Whilst descending, air-brakes are in operation as no steam is used on the downward journey.

During the journey from the summit the chimney lids on the engines are in the closed position. Just before the loco arrives at Llanberis both the lid and the firebox are opened so as to get the necessary amount of steam for the return journey.

Invariably each driver drives the same engine every day and has the services of the same fireman, if possible. On the journey to the summit the guard, sitting in front, keeps a constant look out for any restriction on the line such as stones and on the return trip this duty is carried out by the fireman. Conversion to oil firing has been a topic of conversation for some time but an experimental period some years ago did not prove very successful and the idea has since been abandoned.

Ralph Sadler, *No. 7, arriving at Hebron station.*

The train en route *for the summit of Snowdon crossing the viaduct. Carriages are always pushed to the top of Snowdon.*

DIESEL LOCOMOTIVES

In the summer of 1986 two diesel engines, each costing a quarter of a million pounds, were bought, and after initial teething troubles are now in daily use. They are named NINIAN in honour of Ninian Davies, chairman of the Snowdon Mountain Railway, and of the Davies family who have been associated with the railway since 1922; and YETI, 'dedicated to all creatures of the mountains, living and legendary'. The name 'yeti' was chosen after a competition on one of the national breakfast-time television programmes.

CARRIAGES

The Lancaster Railway Carriage and Wagon Co. Ltd in Lancashire was the company that built the majority of the carriages, though later some were manufactured in Switzerland. The earliest carriages were open with fifty passengers sitting on wooden seats with the guard in front. In 1895 the company purchased five but it eventually agreed to have carriages with side partitions and roofs and some even had curtains to protect the passengers from the wind and rain — even the sun on occasions — and indeed from the smoke of the train itself at times (see photograph on back cover). There was some concern in those early years that glass windows could give too much wind resistance at Clogwyn station and as a result the sides of the carriages were unprotected. Today of course, all carriages have glass windows on the doors and sides. The brake system on all carriages is independent to those on the locomotive.

In later years one of the original carriages was converted for the use of the track maintenance personnel, and is referred to as 'the truck'. It can be seen being loaded in the early morning at Llanberis with provisions for the summit hotel. It has a 700 gallon tank for carrying water to the hotel and a small hut for the permanent way-men to shelter. The 'caboose' as it is sometimes known has its own braking

HOW TO REACH THE SUMMIT OF SNOWDON

(3570 feet above the Sea)

Travel to LLANBERIS, and thence to the actual SUMMIT OF SNOWDON by the . . .

SNOWDON MOUNTAIN "RACK" RAILWAY

The only Mountain Railway in Britain

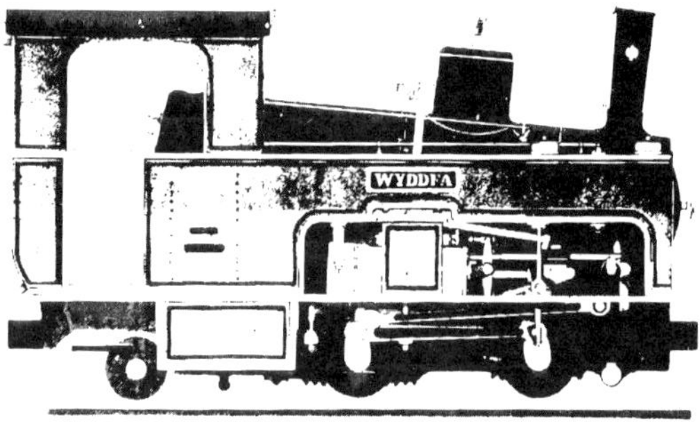

THE SNOWDON MOUNTAIN RAILWAY

TO THE TOP WITHOUT TOIL

An hour to go up and an hour to come down. Magnificent Views en route and from the Summit

SNOWDON SUMMIT HOTEL

is the highest building in Britain . . .

LLANBERIS

lies at the foot of Snowdon, in the heart of Snowdonia

Descriptive Leaflet and Information from Manager, Dept. S.P., Snowdon Railway, Llanberis. Tel. Llanberis 225

(Photograph by courtesy of Reg Hughes.)

The maintenance vehicle, known as the 'caboose' or the 'truck'.

system and one cannot but note its odd shape, caused no doubt by the nine degree angle between it and the flat surface of the original truck.

All the rolling stock is painted in impressive red and yellow with the motif of the company suitably displayed.

SIGNALS

Though non-existent today, signals were in use at the turn of the century as can be seen on many of the postcards printed then (see photograph on back cover). They were installed at the base station, at various places on the journey where two trains could pass each other, and of course at the summit. In 1933 they were removed and today all the inter-communuications between stations is by radio-telephone and the control of trains is based on the 'permissive ticket' system.

The track is of a gauge of 2 feet 7½ inches (800 mm) and the sleepers are of rolled steel inserted at 3 feet apart, though a different type of sleeper has been laid to replace the original ones — it is estimated that 10,000 are in use from the base station to the summit.

STATIONS

The line has five stations though the train at one time used to stop at the Waterfall station as well. The five are the base station at Llanberis, Hebron, Halfway, Clogwyn and the summit itself and there is a small platform at Rocky Valley just below Clogwyn which is the furthest point to which the train will travel if the weather is inclement. Hebron was named because of its proximity to Hebron Chapel. In the early years of this century there was a thriving community in the smallholdings around the chapel and animal foods and other agricultural necessities were brought from Llanberis on the train to Hebron, but this source of income to the railway company did not come up to expectations. The platform at Hebron is one hundred yards long and it was intended to have a second platform but the idea never materialized. There was also a siding at Hebron but it was lifted in 1920. Hebron station is at an altitude of 930 feet above sea level.

There were no station buildings at the Halfway stage at one time but today not only is there a small hut but also a reservior to hold 9,000 gallons of water for the trains. The reservoir is kept filled by a connecting pipe from the overflow at Clogwyn station. Those who walk up along the Snowdon path are glad to rest their weary limbs at the Halfway hut situated below the station. The refreshment hut is not in the occupancy of the railway company but is owned by a family from Llanberis who travel up daily to prepare their delicious lemonade — the preparation of which is kept secret! Halfway station is 1,614 feet above sea level. The station at Clogwyn is situated in the most unprotected part of the whole journey, 2,541 feet up, with a splendid view of the hamlet of Nant Peris and the Llanberis Pass below. Towards the west, and in the distance beyond, can be seen the vast expanse of the rolling mountain slopes and in the distance the flat acres of the Isle of Anglesey. Clogwyn, of course, is near the renowned Clogwyn Du'r Arddu and a wind speed of 34 knots is the operational limit.

The present summit hotel was built in 1936. As previously mentioned, a cluster of wooden huts had been erected at the summit to sell refreshments and to offer some kind of accommodation. During the winter months the hotel is barred and closed. Recently a substantial amount of money was spent on refurbishing the hotel so that the half million people who reach the summit by foot or by train are well catered for. During the harsh winter months, naturally, extensive damage is done to the doors and windows and such cost is accentuated by the desire of some vandals to break in to seek shelter. But not all shelter-seekers are so insensitive — on one occasion someone had broken the lock of the door at the Halfway hut, but had left a sum of money to pay for the damage!

The popularity of the mountain railway has certainly increased over the years — 12,000 passengers were carried in 1897, 76,539 in 1985 and over 120,000 in 1989. Over the years the company has sold its interests in both the Royal Victoria and Padarn Villa hotels.

It has been a tradition in the area that sons follow fathers in the employment of the railway company and some familes can relate some kind of association between all members of the staff with one or two being witty in the extreme. There is a story — true of false is unknown — but it is worth recording. One of the drivers, the late R.A. Jones ('Bobbie Cwcs' to all his acquaintances), is reported to have gone to see the manager at the end of the afternoon to complain about the poor service he was getting from his fireman. The manager, so it seems, was rather reluctant to give Bobbie a replacement. "Well," replied Bobbie, "if you cannot give me another fireman, then get me a smaller mountain!"

Half Way station with Snowdon in the background.
(Photograph by courtesy of Gethin Williams.)

The cutting in the rock near Half Way.
(Photograph by courtesy of Gethin Williams.)

The present hotel on the summit.

The only shelter at Clogwyn station, 2,556 feet above sea level!

(Photograph by courtesy of Gethin Williams.)

Victoria Hotel, 1897

Examples of post cards typical of those sent from the summit in days gone by.

(By courtesy of Iwan Roberts.)

The shop and café near the car park in Llanberis.

Let Comfort Reign

It may be true that if I walked
   The twisting path alone,
I might above the mist espy
   Some chieftain on his throne.

Or maybe gasp at nature's gift
   Whilst lingering by the brook,
Or climb the chapel stile to curse
   One more nostalgic look.

But for today, I'll join the throng,
   Why not? Let comfort reign,
For to the summit I will go
   In peace — by Snowdon train.

Rol Williams

Mr a Mrs Mc Dermid yn derbyn eu gwobrau oddi wrth Derek Rogerson a Graeme Sutherland(ar y dde) rheolwr Gwesty'r Fictoria.

9/89 **Y tymor gorau erioed**

MAE Tren Bach y Wyddfa wedi mwynhau eu tymor mwyaf llwyddiannus erioed.

Ddydd Llun, cyrhaeddodd y gwasanaeth gyfanswm o 100,000 o deithwyr am y flwyddyn - y tro cyntaf iddyn nhw gyrraedd y ffigwr yma.

A derbyniodd y 100,000ed teithiwr, Daniel Mc Dermid o Glasgow, docyn oes ar y rheilffordd

...ailway at peak performance

DESPITE the Welsh weather, the Snowdon Mountain Railway carried 97,578 passengers during 1988, managing director Mr Derek Rogerson wrote in a newsletter to shareholders yesterday.

The total was the third highest in the company's 93-year history – 9.7pc higher than the 1987 total of 88,933, and only slightly less than the record of 99,695 carried in 1975.

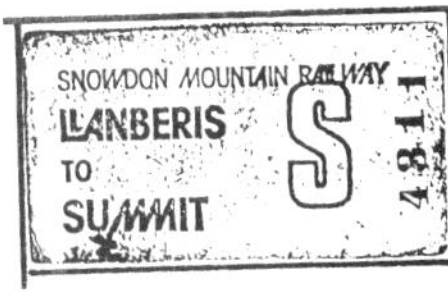

Weather-hit railway 'still on right track'

10 Jan 7/85 1985

by Gerald Williams

SNOWDON Mountain Rail-

...ight—Snowdon mountain railway

...g to the summit, ...on has started ...s many as 13 ...ne day during ...key Valley. ...said that, now

A 125-year ground lease of Caernarfon Airport—the former RAF aerodrome at Llandwrog—has been bought for £140,000, including the price of equipment.

It is the only civil airport ...les, and Mr ...reholders that ...mpany, Snow... Aviation, has ...d to operate

...rs feel that ...to develop the ...ere are plans ...iation museum ...ather tourist

Railway fitter calls it a day

A SNOWDON Mountain railway worker who recall... nightlong sessions of stoking up the old-time steam engine... retires today (Friday)

He is fitter, Mr. Goronw... Roberts, 64, of 7 Warde... Street, Llanberis, who start... ...ng for the compa... ...0 years ago.

...t started with ... ...lyndwr Jones, ...gin the fires in ... ...omotives at 10 pm

...ain climbs to a new record

Going up – the Snowdon Mountain Railway

Nov 1989 DP

Snowdon trip for 120,000

THE Snowdon Mountain Railway carried 8,000 tonnes of people to the top of the highest mountain in England and Wales during the record 1989 season, the best since it opened 94 years ago. The railway is now closed for the winter.

A total 120,826 passengers were carried, beating the previous record of 99,695 set in ...

task for the seven vintage steam locos and two modern diesels.

With the exception of the tracks in the engine shed, there is not an inch of level track on the railway. It is uphill all the way, the steepest section at 1 in 5.5, or 18.2pc, and the vertical elevation between the base station at ...

Other books of interest to railway enthusiasts:

Rheilffyrdd yng Ngwynedd: Ddoe a Heddiw
Railways in Gwynedd: Then and Now
Ken Robinson
Full of black and white photographs of trains, stations and lines.
£2.25

James Gresham and the Vacuum Railway Brake
Colin A. Gresham.
£2.00

Lein Bach y Penrhyn
J.K. Jones
The history of the Penrhyn Quarry railway with numerous black and white photographs.
£1.50

Cyhoeddiadau Mei, Pen-y-groes, Caernarfon, Gwynedd LL54 6NG. Tel./Fax.: (0286) 880302